T0213573

Lecture Notes in Artificial Intelligence   **9549**

Subseries of Lecture Notes in Computer Science

Jeffrey T.K.V. Koh · Belinda J. Dunstan
David Silvera-Tawil · Mari Velonaki (Eds.)

Cultural Robotics

First International Workshop, CR 2015
Held as Part of IEEE RO-MAN 2015
Kobe, Japan, August 31, 2015
Revised Selected Papers

 Springer

Editors
Jeffrey T.K.V. Koh
Creative Robotics Lab, UNSW Art & Design
Sydney
Australia

David Silvera-Tawil
Creative Robotics Lab, UNSW Art & Design
Sydney
Australia

Belinda J. Dunstan
Creative Robotics Lab, UNSW Art & Design
Sydney
Australia

Mari Velonaki
Creative Robotics Lab, UNSW Art & Design
Sydney
Australia

ISSN 0302-9743   ISSN 1611-3349   (electronic)
Lecture Notes in Artificial Intelligence
ISBN 978-3-319-42944-1   ISBN 978-3-319-42945-8   (eBook)
DOI 10.1007/978-3-319-42945-8

Library of Congress Control Number: 2016945942

LNCS Sublibrary: SL7 – Artificial Intelligence

This Springer imprint is published by Springer Nature
The registered company is Springer International Publishing AG Switzerland

Preface

During the development of the IEEE RO-MAN2016 Cultural Robotics Workshop, we discovered an ever-expanding body of projects that could be classified as "cultural robotics"; robotic improvisational jazz musicians, a robot leading morning prayers, robot bartenders and ballet dancers, robots in theatrical performances, and more. It became very clear to us that robots were playing an increasing role in the production of culture, and that this role was collaborative, sincere, and significant. In research and in the media, more examples emerge every day. It is this significance that has motivated us to collate and share the resulting publications of our workshop.

Our call for contributions was answered by over 50 researchers from countries all over the world, including Australia, Egypt, Japan, Peru, Denmark, and Singapore. A total of 12 full papers and one short paper were accepted from 26 initially submitted. The diversity of the papers confirmed that the ways in which robots were shaping, and will continue to shape, human culture was already extending to areas of our lives that we had not imagined. The selected authors demonstrated a commitment to research investigating our key line of inquiry, that is: "What is the future of robotic contribution to human cultures?" Many of them offered unique and critical insights into the surrounding issues of this intersection of technology and culture, including educational, sociological, and gender–political concerns.

In collating the following papers, we hope to contribute in breadth and depth to the field of cultural robotics, and generate further discourse on the questions that emerged from the workshop discussions, including, "What will the advent of robotic-generated culture look like?"

The papers are organized into four categories. These categories are indicative of the extent to which culture has influenced the design or application of the robots involved, and they explore the progressive overlap between human- and robotic-generated culture. These categories are defined and explored in the opening chapter.

We would like to thank our contributing authors for their enthusiasm, commitment, and hard work: Your expertise and generosity in the submissions and discussions engendered an inspiring workshop and publication.

We would like to acknowledge the UNSW NIEA Creative Robotics Lab (CRL), and its director, Associate Professor Mari Velonaki. Thank you for founding a lab that is as close as a family, and inspires a creative approach to robotics research that is being met with international interest and vigor.

June 2016

Jeffrey T.K.V. Koh
Belinda J. Dunstan
David Silvera-Tawil

Organization

Program Committee

Charith Fernando	Keio University, Japan
Hank Haeusler	University of New South Wales, Australia
Guy Hoffman	Cornell University, USA
Benjamin Johnston	University of Technology, Sydney, Australia
Yoshio Matsumoto	National Institute of Advanced Industrial Science, Japan
Ryohei Nakatsu	National University of Singapore, Singapore
Roshan Peiris	National University of Singapore, Singapore
Doros Polydorou	University of Hertfordshire, UK
David Rye	The University of Sydney, Australia
Maha Salem	Google, UK
Hooman Samani	National Taipei University, Taiwan
Dag Sverre	University of Hertfordshire, UK
James Teh	National University of Singapore, Singapore
Katsumi Watanabe	Waseda University, Japan
Mary-Anne Williams	University of Technology, Sydney, Australia
Kening Zhu	City University of Hong Kong, Hong Kong, SAR China

Contents

Robots as Producers of Culture: Material and Non-material

The Advent of Robotic Culture

Introduction

Cultural Robotics: Robots as Participants and Creators of Culture

Belinda J. Dunstan$^{(\boxtimes)}$, David Silvera-Tawil, Jeffrey T.K.V. Koh, and Mari Velonaki

Creative Robotics Lab, The University of New South Wales, Sydney, Australia
{belinda.dunstan,d.silverat,jeffrey.koh,mari.velonaki}@unsw.edu.au
http://www.crl.niea.unsw.edu.au

Abstract. This introductory chapter reviews the emergence, classification, and contemporary examples of cultural robots: social robots that are shaped by, producers of, or participants in culture. We review the emergence of social robotics as a field, and then track early references to the terminology and key lines of inquiry of *Cultural Robotics*. Four categories of the integration of culture with robotics are outlined; and the content of the contributing chapters following this introductory chapter are summarised within these categories.

Keywords: Cultural robotics · Social robotics · Human-robot interaction · Culture

1 Introduction

Over the last decade the field of robotics has seen a significant increase in human-robot interaction (HRI) research [16]. It is expected that the next generation of robots will interact with humans to a much greater extent than ever before [18]. As the distance between humans and robots narrows, robotics research is moving into areas where robotic entities have become participants, and in some cases generators of culture. With this in mind, this introductory chapter aims to identify and outline the notion of *Cultural Robotics* as an emerging field.

As a logical evolution from the field of HRI, and particularly social robotics, the emerging field of cultural robotics aims to understand the role of robots as cultural participants and creators of culture [11,35]. Cultural robotics, then, is defined as the study of robots in culture, cultural acceptance of robots, and robot-generated culture. In other words, a cultural robot is a robotic entity that participates in, and contributes to, the development of material and/or non-material culture. The terms 'material' and 'non-material' refer to tangible cultural products such as a painting or a piece of music and intangible products such as values, norms and traditions respectively.

In this vein, previous research has investigated the effect of culture on both the design [25] and acceptability of robots [1,17]. The cultural influence of

© Springer International Publishing Switzerland 2016
J.T.K.V. Koh et al. (Eds.): Cultural Robotics 2015, LNAI 9549, pp. 3–13, 2016.
DOI: 10.1007/978-3-319-42945-8_1

robots, furthermore, can be noticed in theatre [27], stand-up comedy [21], interactive art [8] and religion [31]; all activities previously reserved for humans but today also 'enjoyed' by robots. Additionally, robots such as "Shimon" [19]—an autonomous robotic marimba player—already produce material cultural artefacts, such as an improvised piece of music. Shimon not only communicates a meaningful message to the human counterpart through shared conventions of communication and musical composition, but in turn provides a new avenue for human/robot collaboration that could lead towards a new musical genre. In addition to serving beverages, the "Robot Bartender" [12] recognises and interprets multimodal socio-cultural signals from its human 'clients'.

As noted, culture is a multilayered construct inclusive of not only external artefacts such as language and customs, food and dance, but also nuanced elements of "a group's shared set of specific basic beliefs, values, practices and artefacts that are formed and retained over a long period of time" [38]. We recognise culture as a complex and integral consideration in the design, application and advancement of social robotics. In looking at the social acceptance of robots, we present culture as the touchstone for meaningful and sustained human-robot interaction.

This chapter will present a survey of all aspects of cultural robotics, such as the impact of culture in the design and application of robots, the cultural acceptance of robots and the emergence of robot-generated culture. It serves as an introduction to the book's contributed chapters from a cross-disciplinary group of researchers and practitioners from fields such as HRI, engineering, computer sciences, art and design.

2 Social Robotics as the Foundation to Cultural Robotics

When the field of *Social Robotics* was first introduced it was applied to multi-robot systems inspired by the collective behaviours of birds, insects, fish or any creature within the same context [4]. With time, the term changed to study a more heterogeneous group which involves the interaction between humans and robots, particularly robots that people anthropomorphize in order to interact with them.

Despite the fact that robots are not sociable in the way humans are [20], in the early stages of social robotics (as known today) some researchers used the term *sociable robots* to distinguish between the more recent human-robot interaction and earlier work [4]. Robot designers, however, tend to use similar social models to those used during human-to-human interaction to incorporate robots into social environments. Although the robots are not strictly 'sociable', the social models they use are based on the foundation of human interaction and, when interacting with robots, the same social models are unconsciously applied by humans [4].

Although the term social robotics is now widely accepted, some researchers argue about the difficulty of creating social robots without a clear understanding of conscience [33]. They point out that morality and ethics are inherent to an individual, defined by his or her relationship to others, and not easy to program or emulate into a robot. Roboticists and social researchers are also beginning to appreciate the importance of social, emotional and ethical issues raised by the development of robots. For example, there has been work on social and moral relationships [10,20,28,41]; the concept of 'personal space' [42]; radical-uncertainty [37]; free choice [2]; self-consciousness [5]; and long-term social interaction [3,15,36] between humans and robots.

The field of social robotics, today, is concerned with the study of all forms of human-robot interaction within a social context, including the appearance and behaviour of socially interactive robots. To different degrees, all social interactions are culturally driven. As a response to the significant growth in social robotics, the field of cultural robotics was recently introduced.

3   Background to Cultural Robotics

For some time now there has been research directed towards building robots that can interact with humans in a social and culturally meaningful way. In fact, some of the earliest examples of embodied robotic agents were in cultural applications, including a radio-controlled anthropomorphic robot titled "K-456" by Nam June Paik and Shuya Abe built in 1964. K-456 was a provocative and controversial political piece; an androgyne in terms of gender identity, the robot played a recording of John F. Kennedy's inaugural address and excreted beans. In 1970 Edward Ihnatowicz produced "Senster," the first robotic sculpture controlled by a digital computer. This large scale interactive system was responsive to sound and low level movement, but would shy away from loud sounds and violent movements, encouraging the audience to adapt their behaviour in an affective response to the movements of the robot.

In spite these early examples, within the traditional robotics community the question of culture has been primarily considered in relation to the reception of robots or the level of general technology acceptance within a particular culture. A number of studies have been conducted comparing preferences between different countries and cultures, for design factors including the size, capability, intelligence and 'life-likeness' of social robots [1,17,25]. However, within the last six years, a new conversation concerning the role of cultural considerations within robotics has emerged.

In 2010, in response to the prevailing linear 'technologically deterministic' scientific discourses on social impacts and acceptability of robotics, Selma Šabanović proposed a framework for 'bi-directional shaping' between robotics and society. Šabanović's publication "Robots in Society, Society in Robots" [40] focuses on "analysing how social and cultural factors influence the way technologies are designed, used, and evaluated as well as how technologies affect our construction of social values and meanings" [40, p. 439]. Together with the observations of MacKenzie and Wajcman [26], Šabanović identifies an existing linear and technocentric trend in technology research and acceptance where "society fills a passive role" and the public is encouraged to view technological change as inevitable and "adapt to technology... not shape it" [26, p. 5]. Further, she notes that despite the significant social implications of robotics research, society is often not included in the design process until the final testing and evaluation stages [40, p. 440].

With a desire to address the nature of this dynamic, Šabanović proposed an approach to design which is value-centred, "consciously incorporating social and cultural meaning into design" [40, p. 445]. Her framework is not presented as direct design recommendations, but rather as recommendation for a relationship—or co-production—between society and technology, as one of "continuous feedback between practice, sense-making and design" [40, p. 445]. It is this very desire to address "the role that social-cultural norms, values and assumptions play in the daily practices of designing robotic technologies" [40, p. 440], that has led to the further development of research on the topic of cultural robotics.

The use of the term *Cultural Robotics* was first explored in depth by Hooman Samani et al. [35], who attributed the development of culture in robotics to the cultural values of the designers, the importance of embodiment in robotics, and the current (and potential) learning capacity of robots. Samani et al. proposed the potential progression of robots from simple tools, to luxury items, to members of human society and projected that they would one day become an integral part of our culture, and perhaps develop their own unique culture. Samani et al. discuss culture from a number of angles including the influence of popular culture and media on robot design and acceptance, and the potential use of robots as telepresence technology. They argue that the design and use of robots ought to be informed by a specific cultural context, and used as both a product and a medium to contribute to the sustainability of cultural practices.

In response and addition to this research, Dunstan and Koh [11] published on the emergence of cultural robotics, defining it as "the study of robots in culture, cultural acceptance of robots, robot-cultural interaction and robot-generated cultures" [11, p. 134], and a social robot as one that contributes to the generation of material and non-material culture. Here, Dunstan and Koh outlined stages of cultural interaction, moving beyond the influence of the values of the designers to identifying specific cases through three stages of immersion; firstly, a robot as an actor within a particular culture; secondly, a robot as a participant in or producer of culture; and thirdly, the potential for the advent of robotic community culture. By surveying emerging social robotics projects from non-traditional robotics conferences, together with analysis of cultural determinants within a cognitive behavioural model, they predicted an increasing integration of culture in robotics and robotics in culture.

The papers within this publication demonstrate that the extent to which human and robot culture overlap and intertwine is now reaching well into the category of 'robot-generated culture' as robots are used to teach, plan and lead culturally meaningful activities. Robots are also generating a new branch of cultural and philosophical inquiry into the roles of gender, embodiment, ethics, performance, and politics in technology.

4   Latest Work in Cultural Robotics

This section presents an overview of the latest works in Cultural Robotics, as exemplified through the submissions to this publication. The submissions are divided into four sections, demonstrating the layers of integration of culture in robotics, and robotics in culture. Namely, these are: (1) culture affecting the design, application and evaluation of robots, (2) robots as participants in culture, (3) robots as producers of culture: material and non-material, and (4) the advent of robotic culture. The following is an overview of each section, and the chapters included therein.

4.1   Culture Affecting the Design, Application and Evaluation of Robots

As mentioned in Sect. 3, in the robotics community the question of culture has been primarily considered in relation to the design and evaluation of socially interactive robots as perceived by a cross-cultural population. In this vein, Yasser Mohammad and Toyoaki Nishida [30] present, in Chapter Two, a comprehensive review of cross-cultural differences in the perception of robots, and include results from an experiment that investigate cross-cultural changes in robot perception using the back-imitation effect, where participants from different cultural backgrounds are required to imitate a robot's behaviour.

Then, in Chapter Three, Hyelip Lee et al. [24] introduce the process followed to design M4K, a telepresence robot created in response to globalisation and the need of people to communicate, and interact, across distance. This robot exceeds the common capacity of bi-directional communication by integrating the ability of tele-manipulation. In this chapter the authors present the main considerations followed during the robot design, considering not only the environment where it will be placed and the tasks that it should achieve, but also the robot's appearance and behaviour that would improve its social acceptability. In this case, the robot would be used as an extension of a user rather than as an individual, independent agent.

In Chapter Four, furthermore, Mauricio Reyes et al. [34] explore the use of a robot's facial expressions during collaborative tasks with humans. Facial expressions, strongly affected by social and cultural context, play a significant role during the communication and interpretation of emotions. This chapter investigates, particularly, the effects of negative facial expression feedback (i.e. sadness) communicated by a robot during a failed human-robot collaborative task, and investigate if human intervention exists on the initial presence of an unexpected failure, and how the intervention is affected by the robot's facial expression.

Clearly, the evaluation of human behaviour and robot perception in a social, cross-cultural environment is complex, and significant work is still needed. In Chapter Five, Diego Compagna et al. [7] introduce a sociology-based theory-driven method to evaluate HRI, and identify aspects of successful and satisfying interactions. The method is based on "a definition of social interaction based on the symbolic interactionism paradigm."

4.2   Robots as Participants in Culture

The participation and integration of robots in culture is demonstrated in Chapter Six with a study conducted by Evgenios Vlachos et al. [39], which aims to provide insight on how users communicate with an android robot and how to design meaningful human robot social interaction for real life situations. The study was initially focused on head orientation behaviour of users in short-term dyadic interactions with an android, however, the results of this study revealed unexpected findings: the female participants spent a significantly longer time interacting with the robot, and further, the setting of an art gallery proved to be a rich context for measuring human-robot interaction. This chapter observes diversities in human-robot interaction behaviour between groups and individuals, and between genders, and most compellingly, that as robots are moved out of the laboratory and into a cultural setting, their reception and the behaviour of participants interacting with them changes in unanticipated ways.

From the art gallery to the classroom, in Chapter Seven, Christian Penaloza et al. [32] discuss their research that explores the potential use of robots as educational tools for non-technology related fields such as history. The authors explore this unique application of robots not only as a means to engage the attention of students, but as a methodological approach for designing the morphology of educational robots, inspired by the ancient gods and historical characters of South American cultures. This chapter includes a number of conceptual designs for culturally-inspired robot morphologies, and cultural educational activities centred around building a robot.

As demonstrated in Chapter Eight through the work of Petra Gemeinboeck and Rob Saunders [14], not only are we seeing the emergence of robot participation in culture, but increasingly, the use of cultural activities to shape the morphology and movement planning of social robots. In this chapter the authors discuss a novel approach towards socializing non-anthropomorphic robots, which involves the 'Performative Body Mapping' of the movement of dancers, to teach non-humanlike robots to move in affective and expressive ways. The authors conduct a number of experiments that attest to the potential of movement to turn an abstract object into an expressive, empathy inducing social actor.

The inclusions of robots in cultural settings generates a number of new questions and discourses. In Chapter Nine, the question of subjectivity and objectivity in films and visual culture is discussed, as increasingly, the use of robotic camera systems removes the human operator entirely from the production and interpretation of images and film. Author Chris Chesher [6] discusses the use of motion control systems and robotically-controlled cameras, and how these alter image genres, and question the audience's perception of subjectivity, surveillance, intimacy, and the uncanny.

Within cultural contexts, we see that the applications of robots are moving beyond the role of 'servant' or worker simply performing efficient assembly-line tasks, but rather, are increasingly involved in creative activities. In Chapter Ten, Christian Laursen et al. [23] discuss the way in which robots can not only support, but spark the imagination of dessert chefs working in food preparation and plating. The authors present a range of prototypes that explore robots providing a role in the creation of aesthetic interactions and experiences regarding the preparation, serving and consumption of food. This research not only presents robots as participants in a culturally rich environment (the kitchen), but even more significantly, it demonstrates the ways in which robots can support and enhance human creativity and move towards being classified as producers of culture.

4.3 Robots as Producers of Culture: Material and Non-material

Since the 19th century, robots have played an important role not only as participants, but also as producers of culture. Early examples include the use of dummies and mechanical puppets: *Automata* (Ernst T.A. Hoffmann, 1814) and *The Sandman* (Ernst T.A. Hoffmann, 1817). Popular media, furthermore, have used robots to create a vision of what the future could be, with human-looking robots contributing and interacting with people as 'equals': *The Bicentennial Man* (Isaac Asimov, 1976). Although we are still far from this impression, in Chap. 11 Elena Knox [22] presents Geminoid-F, a female-appearing Android robot, as the main character of an experimental video artwork—*Comfortable and Alive*—created to facilitate a wider, yet fractional discussion of the cultural provenance and potential integration of female-appearing robots.

From cinema to the performing arts, through the work of Wade Marynowsky et al. [29], Chap. 12 shows how framing a robot-based performance as a Gesamtkunstwerk—a work that synthesizes all art forms—contributes to the creation of culture. In this chapter Marynowsky et al. present "Robot Opera" and the history and exploration of robots in the performing arts. Following a similar direction, in Chap. 13 Petra Gemeinboeck and Rob Saunders continue the discourse of the cultural legacy of robots in the performing arts [13], including historical and contemporary works that explore the 'machine creativity' as a cultural, bodily practice, where machines (robots) are performers capable of expanding the 'script' given by their human creators.

4.4 The Advent of Robotic Culture

In this final section we explore the advent of robotic culture, through the work of Alex Davies and Alexandra Crosby [9], in Chap. 14. In this chapter the authors present the 'on-stage' and 'off-stage' storyworld of the first all-robot band, Compressorhead. Here the authors argue that robots can indeed be seen not only as performers, but even as celebrities and therefore be taken seriously as participants and producers of material (e.g. music and merchandise) and non-material (e.g. social values and norms) culture, and further, they point towards the real emergence of autonomous robotic-generated culture.

5   Conclusions and Future Direction

At the RO-MAN 2015 conference, we were so fascinated to watch short films presented by the authors of robots so deeply immersed in cultural practices;

robots being carefully dressed in traditional robes by children who were being taught about ancient cultures (by the robots!); robots gently spiralling chocolate to assist a dessert chef with plating a dish; and a human dancer in a large geometric costume, mapping fluid human gestures for robotic movement planning. Reflecting on our key line of inquiry, 'What is the future of robotic contribution to human cultures?', while the answer grows and changes almost daily, the nature of the contribution is emerging; one which is substantial, considered, nuanced, and deeply significant.

As technology advances, we believe that the role of robots will change from interactive social agents with the ability to emulate and respond with human-like social behaviours, to independent, emotional and intellectual entities with the ability to create their own identity. For this to happen, however, significant work is needed. To date, most socially interactive robots don't have the ability to work unattended, for extended periods of time, without human intervention. In fact, most social robots (if not all of them) are either remotely operated or follow a very specific set of rules that define their social/cultural behaviour. Technological advances in artificial intelligence will allow robots to have their own 'intelligence,' learn and make independent decisions, creating a world of opportunities for them to participate and create their own culture. Through this ability, we believe, continuously-evolving socially-interactive robots that adapt to human behaviour will be created.

Currently, interaction with a social robot is still something most people only experience as part of an experiment or on a very rare public occasion. In order to gain a deeper understanding of the interaction capacity and potential use of social robots in cultural settings, more robots need to be moved out of the laboratory and into art galleries, kitchens, classrooms etc.; where the benefit of their inclusion in these settings, for both testing and participation, are illustrated clearly by the contributions to this publication.

We hope to continue to contribute to the conversation around the emergence of robot generated culture, and we anticipate that this will be the category of cultural robotics which will see the most rapid and interesting growth in the next few years.

References

1. Bartneck, C.: Who like androids more: Japanese or US Americans? In: Proceedings of IEEE International Symposium on Robot and Human Interactive Communication (2008)
2. Bello, P., Licato, J., Bringsjord, S.: Constraints on freely chosen action for moral robots: consciousness and control. In: Proceedings of IEEE International Symposium on Robot and Human Interactive, Communication, pp. 505–510 (2015)
3. Bickmore, T.W., Picard, R.W.: Establishing and maintaining long-term human-computer relationships. ACM Trans. Comput.-Hum. Interact. **12**(2), 293–327 (2005)
4. Breazeal, C.: Toward sociable robots. Robot. Auton. Syst. **42**, 167–175 (2003)

5. Bringsjord, S., Licato, J., Govindarajulu, N.S., Ghosh, R., Sen, A.: Real robots that pass human tests of self-consciousness. In: Proceedings of IEEE International Symposium on Robot and Human Interactive, Communication, pp. 498–504 (2015)
6. Chesher, C.: Robots and the moving camera in cinema, television and digital media. In: Koh, J.T.K.V., Dunstan, B.J., Silvera-Tawil, D., Velonaki, M. (eds.) Cultural Robotics. LNAI, vol. 9549. Springer, Switzerland (2016)
7. Compagna, D., Marquardt, M., Boblan, I.: Introducing a methodological approach to evaluate HRI from a genuine sociological point of view. In: Koh, J.T.K.V., Dunstan, B.J., Silvera-Tawil, D., Velonaki, M. (eds.) Cultural Robotics. LNAI, vol. 9549. Springer, Switzerland (2016)
8. Silvera-Tawil, D., Velonaki, M., Rye, D.: Human-robot interaction with humanoid diamandini using an open experimentation method. In: Proceedings of IEEE International Symposium on Robot and Human Interactive, Communication, pp. 425–430 (2015)
9. Davies, A., Crosby, A.: Compressorhead: the robot band and its transmedia storyworld. In: Koh, J.T.K.V., Dunstan, B.J., Silvera-Tawil, D., Velonaki, M. (eds.) Cultural Robotics. LNAI, vol. 9549. Springer, Switzerland (2016)
10. Duffy, B.R.: Anthropomorphism and the social robot. Robot. Auton. Syst. **42**, 177–190 (2003)
11. Dunstan, B.J., Koh, J.T.K.V.: Cognitive Robotics, Chapter A Cognitive Model for Human Willingness to Collaborate with Robots: The Emergence of Cultural Robotics, pp. 127–142. CRC Press, Boca Raton (2015)
12. Foster, M.E., Gaschler, A., Giuliani, M.: How can I help you?: comparing engagement classification strategies for a robot bartender. In: Proceedings of ACM International Conference on Multimodal Interaction, pp. 255–262 (2013)
13. Gemeinboeck, P., Saunders, R.: The performance of creative machines. In: Koh, J.T.K.V., Dunstan, B.J., Silvera-Tawil, D., Velonaki, M. (eds.) Cultural Robotics. LNAI, vol. 9549. Springer, Switzerland (2016)
14. Gemeinboeck, P., Saunders, R.: Towards socializing non-anthropomorphic robots by harnessing dancers kinesthetic awareness. In: Koh, J.T.K.V., Dunstan, B.J., Silvera-Tawil, D., Velonaki, M. (eds.) Cultural Robotics. LNAI, vol. 9549. Springer, Switzerland (2016)
15. Gockley, R., Bruce, A., Forlizzi, J., Michalowski, M., Mundell, A., Rosenthal, S., Sellner, B., Simmons, R., Snipes, K., Schultz, A.C., Wang, J.: Designing robots for long-term social interaction. In: Proceedings of IEEE/RSJ International Conference on Intelligent Robots and Systems, pp. 1338–1343 (2005)
16. Goodrich, M.A., Schultz, A.C.: Human-robot interaction: a survey. Found. Trends Hum.-Comput. Interact. **1**(3), 203–275 (2007)
17. Haring, K.S., Silvera-Tawil, D., Matsumoto, Y., Velonaki, M., Watanabe, K.: Perception of an android robot in Japan and Australia: a cross-cultural comparison. In: Beetz, M., Johnston, B., Williams, M.-A. (eds.) ICSR 2014. LNCS, vol. 8755, pp. 166–175. Springer, Heidelberg (2014)
18. Harper, C., Virk, G.: Towards the development of international safety standards for human robot interaction. Int. J. Soc. Robot. **2**, 229–234 (2010)
19. Hoffman, G., Weinberg, G.: Shimon: an interactive improvisational robotic marimba player. In: Proceedings of ACM Conference on Human Factors in Computing Systems (2010)
20. Kahn Jr., P.H., Freier, N.G., Friedman, B., Severson, R.L., Feldman, E.N.: Social and moral relationships with robotic others? In: Proceedinhs of IEEE International Workshop on Robot and Human Interactive Communication, pp. 545–550 (2004)

21. Katevas, K., Healey, P.G., Harris, M.T.: Robot stand-up: engineering a comic performance. In: Proceedings of Workshop on Humanoid Robots and Creativity at the IEEE-RAS International Conference on Humanoid Robots (2014)
22. Knox, E.: 'Face robots' on screen: comfortable and alive. In: Koh, J.T.K.V., Dunstan, B.J., Silvera-Tawil, D., Velonaki, M. (eds.) Cultural Robotics. LNAI, vol. 9549. Springer, Switzerland (2016)
23. Laursen, C., Pedersen, S., Merritt, T., Caprani, O.: Robot-supported food experiences exploring aesthetic plating with design prototypes. In: Koh, J.T.K.V., Dunstan, B.J., Silvera-Tawil, D., Velonaki, M. (eds.) Cultural Robotics. LNAI, vol. 9549. Springer, Switzerland (2016)
24. Lee, H., Kim, Y.H., Lee, K.K., Yoon, D.K., You, B.J.: Designing the appearance of a telepresence robot, M4K: a case study. In: Koh, J.T.K.V., Dunstan, B.J., Silvera-Tawil, D., Velonaki, M. (eds.) Cultural Robotics. LNAI, vol. 9549. Springer, Switzerland (2016)
25. Lee, H.R., Šabanović, S.: Culturally variable preferences for robot design and use in South Korea, Turkey, and the United States. In: Proceedings of ACM/IEEE International Conference on Human-Robot Interaction (2014)
26. MacKenzie, D., Wajcman, J.: Introductory essay: the social shaping of technology. In: The Social Shaping of Technology, pp. 3–27. Open University Press (1999)
27. Malik, B.: Savanna: a possible landscape (2013). http://www.wallpaper.com/design/ savanna-a-possible-landscape-byamit-drori-and-dover-cederbaum?iid=sr-link1
28. Malle, B.F., Scheutz, M.: When will people regard robots as morally competent social partners? In: Proceedings of IEEE International Symposium on Robot and Human Interactive, Communication, pp. 486–491 (2015)
29. Marynowsky, W., Knowles, J., Frost, A.: Robot opera: a Gesamtkunstwerk for the 21st century. In: Koh, J.T.K.V., Dunstan, B.J., Silvera-Tawil, D., Velonaki, M. (eds.) Cultural Robotics. LNAI, vol. 9549. Springer, Switzerland (2016)
30. Mohammad, Y., Nishida, T.: Cultural difference in back-imitations effect on the perception of robots imitative performance. In: Koh, J.T.K.V., Dunstan, B.J., Silvera-Tawil, D., Velonaki, M. (eds.) Cultural Robotics. LNAI, vol. 9549. Springer, Switzerland (2016)
31. RT News: Iran teacher builds robot to teach children how to pray (2014). http://rt.com/news/iran-praying-robot-children-888/
32. Penaloza, C., Lucho, C., Cuellar, F.: Towards the design of robots inspired in ancient cultures as educational tools. In: Koh, J.T.K.V., Dunstan, B.J., Silvera-Tawil, D., Velonaki, M. (eds.) Cultural Robotics. LNAI, vol. 9549. Springer, Switzerland (2016)
33. Ramey, C.H.: Conscience as a design benchmark for social robots. In: Proceedings of IEEE International Workshop on Robot and Human Interactive, Communication, pp. 486–491 (2006)
34. Reyes, M., Meza, I., Pineda, L.A.: The positive effect of negative feedback in HRI using a facial expression robot. In: Koh, J.T.K.V., Dunstan, B.J., Silvera-Tawil, D., Velonaki, M. (eds.) Cultural Robotics. LNAI, vol. 9549. Springer, Switzerland (2016)
35. Samani, H., Saadatian, E., Pang, N., Polydorou, D., Fernando, O.N.N., Nakatsu, R., Koh, J.T.K.V.: Cultural robotics: the culture of robotics and robotics culture. Int. J. Adv. Robot. Syst. 10(400), 1–10 (2013)
36. Shibata, T., Kawaguchi, Y., Wada, K.: Investigation on people living with seal robot at home. Int. J. Soc. Robot. 4(1), 53–63 (2012). doi:10.1007/s12369-011-0111-1

37. Silvera-Tawil, D., Garbutt, M.: The far side of the uncanny valley: 'Healthy persons', androids, and radical uncertainty. In: Proceedings of IEEE International Symposium on Robot and Human Interactive, Communication, pp. 740–745 (2015)
38. Vas, T., Rowney, J., Steel, P.: Half a century of measuring culture: review of approaches, challenges, and limitations based on the analysis of 121 instruments for quantifying culture. J. Int. Manag. **15**(4), 357–373 (2009)
39. Vlachos, E., Jochum, E., Schärfe, H.: Head orientation behavior of users and durations in playful open-ended interactions with an android robot. In: Koh, J.T.K.V., Dunstan, B.J., Silvera-Tawil, D., Velonaki, M. (eds.) Cultural Robotics. LNAI, vol. 9549. Springer, Switzerland (2016)
40. Šabanović, S.: Robots in society, society in robots. Int. J. Soc. Robot. **2**(4), 439–450 (2010). ISSN 1875–4805
41. Wagner, J.J., Van der Loos, H.F.M., Leifer, L.J.: Construction of social relationships between user and robot. Robot. Auton. Syst. **31**(3), 185–191 (2000)
42. Walters, M.L., Dautenhahn, M.L., Boekhorst, R., Koay, K.L., Kaouri, C., Woods, S., Nehaniv, C., Lee, D., Werry, I.: The influence of subjects' personality traits on personal spatial zones in a human-robot interaction experiment. In: Proceedings of IEEE International Workshop on Robot and Human Interactive Communication, pp. 347–352 (2005)

Culture Affecting the Design, Application and Evaluation of Robots

Cultural Difference in Back-Imitation's Effect on the Perception of Robot's Imitative Performance

Yasser Mohammad[1(✉)] and Toyoaki Nishida[2]

[1] Assiut University, Asyut, Egypt
yasserm@aun.edu.eg
[2] Kyoto University, Kyoto, Japan
nishida@i.kyoto-u.ac.jp

Abstract. Cultural differences have been documented in different aspects of perception of robots as well as understanding of their behavior. A different line of research in developmental psychology has established a major role for imitation in skill transfer and emergence of culture. This study is a preliminary cross–cultural exploration of the effect of imitating the robot (back imitation) on human's perception of robot's imitative skill. In previous research, we have shown that engagement in back imitation with a NAO humanoid robot, results in increased perception of robot's imitative skill, human–likeness of motion, and willingness of future interaction with the robot. This previous work mostly used Japanese university students. In this paper, we report the results of conducting the same study with subjects of two cultures: Japanese and Egyptian university students. The first finding of the study is that the two cultures have widely different expectations of the robot and interaction with it and that some of these differences are significantly reduced after the interaction. The second finding is that Japanese students tended to attribute higher imitation skill and human likeness to the robot they imitated while Egyptian students did not show such tendency. The paper discusses these findings in light of known differences between the two cultures and analyzes the role of expectation in the differences found.

1 Introduction

Attitude toward robots is one of the major factors determining the success or failure of future social robots that are expected to occupy our homes, offices, hospitals and schools. One important factor that affects these attitudes is culture.

Culture is a multifaceted and complex concept that may have different meanings for different researchers [19]. In this work, we follow Samani et al. [19] and Taras et al. [21] and define culture as a group's shared set of specific basic beliefs, values, practices and artefacts that are formed and retained over a long period of time. This includes communicative aspects (e.g. nonverbal behaviors including gestures and proximities).

© Springer International Publishing Switzerland 2016
J.T.K.V. Koh et al. (Eds.): Cultural Robotics 2015, LNAI 9549, pp. 17–32, 2016.
DOI: 10.1007/978-3-319-42945-8_2

Previous studies have shown that culture plays an important role in shaping people's attitudes toward robots in several contexts. For example, Bartneck [1] studied the perception of robot anthropomorphism and likability for United States and Japanese subjects and found that Japanese subjects tended to like conventional robots more than US subjects while the reverse was observed for androids (e.g. robots with highly human–like appearance covered with artificial skin) [1]. Finding differences between eastern and western cultures in cross–cultural HRI research is common. Lee and Sabanović [9] studied the acceptability of different robot designs (appearance) by subjects from Turkey, South Korea, and United States. They found that religious belief and media exposure are not enough to explain the discovered differences between people from these countries in their preferences which suggests a specific role of culture. Both of these studies involved measuring people's response to robot representations (e.g. images) rather than actual interactions with them.

It is commonly held that westerners perceive robots differently than easterners because of the difference of their portray in media. A common example is comparing "The Terminator" with "Astro Boy". While the first is a killing machine the later is a helping child–like robot with human–like curiosity and emotions. This conception though is challenged by some research findings. For example Bartneck et al. compared Dutch, Chinese, German, Mexican, American (USA) and Japanese participants based on the Negative Attitude towards Robots Scale (NARS) and found no particularly positive attitudes for Japanese participants [2]. Wang et al. found that Chinese participants expressed more negative attitudes toward robots than American participants [23]. Shibata et al. reported no difference between UK and Japanese participants when subjectively reporting about a Paro robot and found in both cases that physical interaction improves subjective evaluations of the robot [20]. These results taken together does not support the simplistic commonly held belief that eastern people are more accepting of robots than their western counterparts but shows a complicated interaction between several factors including appearance, culture, interaction quality, etc.

Cultural transfer may be mediated by imitation. Nielsen [14] argues that emergence of imitation and play in children was a precursor for the emergence of culture as a complex construct in human life. Imitation is not always a conscious process in humans. For example, Chartrand and Bargh experimentally showed that behavioral mimicry has a significant effect on the interaction and increases empathy towards the interaction partner [3] which is usually referred to as the "chameleon effect". Several HRI studies looked for similar effects when people interact with robots. Riek et al. showed that real–time head gesture mimicry improves rapport between a human and a robot [16].

HRI studies of imitation have focused on the effect of robot's imitative ability on human's perception of the traits of this robot and convincingly argued for a positive effect [16]. In a series of previous studies [11–13], we investigated the opposite case in which a human imitates the robot. The main hypothesis was that this form of back-imitation will have positive effects on the perception of robot's imitative skill and may also lead to more acceptance [11]. We found that back-imitation leads indeed to increased perception of robot's imitative skill and human–likeness of motion and may lead to increased intention of future interaction with it [13]. For the purposes of this study we define back imitation following Mohammad and Nishida [13] as *the imitation of the learner by the teacher during, before or after the demonstration of a new task.*

These studies were conducted using mostly Japanese university student participants and no cultural evaluation was conducted. In this paper, we repeat one of these experiments with participants from Japan and Egypt and show that the positive effects of back-imitation were lacking in Egyptian subjects. We discuss this results in terms of the effect of prior expectation and cultural aspects.

A few studies reported the response of Egyptian subjects to robots. For example, Trovato et al. [22] compared the response of Egyptian and Japanese subjects to a humanoid robot speaking in Arabic (native language of Egypt) and Japanese and found that people from each nationality preferred robots that spoke in their native language and used the culture-specific greeting gestures. The experiment was conducted using only videos of the robot. One problem of this study is that the effect of language understanding may overshadow other cultural differences. Salem et al. [18] conducted a cross–cultural study in which a humanoid robot (Ibn Sina) was displayed in a major exhibition (Dubai's GITEX) and compared the response of people from different nationalities including African Arabs and South eastern Asians. The study focused on the order of robot applications and found significant interplay between religion, age and cultural origin and acceptance of robots in different applications.

This work differs from the aforementioned studies in that it focuses on actual interaction with the robot (a NAO humanoid robot in our case) and measures the effect of a behavioral aspect of the robot instead of its appearance or design. We believe that behavior and motion are as important as appearance in attribution of skill and human–likeness and in general acceptance of the robot for different roles.

Imitative skill in this paper is defined as the objective accuracy in copying limb motions demonstrated by the human. As such, it is related to motion human–likeness which describes the degree by which motion trajectories of robot limbs resemble human motion in general not necessarily the demonstrated behavior. For example, a robot that closes its hand during demonstrating a waving gesture will have low imitative skill but the motion can still be human–like in the sense that it is similar in form to normal human motion in terms of smoothness and respecting human joint range limits. A concept related to human–likeness that we discuss later in this paper is humanness which is defined as the degree by which humanity is ascribed to an agent [5]. Our previous studies found that two factors contribute to this overall assessment of humanness clustering positive traits (e.g. curiousity, sociability, friendliness) and negative traits (e.g. jeouleousy, impatience, distractibility) [13]. These two clusters of features consitute the positive and negative humanness scores in this study. Interaction quality is defined here as the participant's overall subjective evaluation of her interaction with the robot.

The rest of the paper is organized as follows: Sect. 2 details the experimental design used in this study and comments on different design choices. Section 3 reports the results of the study and Sect. 4 discusses their implications. The paper is then concluded.

2   Experimental Design

The design of this experiment is similar to the main study reported in [13]. The main difference is that participants came from two different nationalities (Egyptian and Japanese). This entailed employing appropriately different statistical analysis of the questionnaires.

The experiment was conducted in Japan which allowed us to recruit 36 Japanese subjects but only 10 Egyptian subjects. We used the data of only 10 Japanese subjects who participated in the experiment reported in [13] selected to match the gender, age and education level of the 10 available Egyptian participants. This is achieved by removing all female Japanese subjects (as all Egyptian subjects were males), we then removed younger Japanese subjects until we had 15 subject of which we picked 10 subjects randomly. This led to 20 participants in total for this study. All participant were male with average age of 26 years for Japanese participants and 30 years for Egyptian participants. Sixteen of the participants were studying STEM subjects and the other four were majoring in humanities (one from Egypt and three from Japan). It should be noted that we found no difference based on educational background (STEM/humanities) in any of the aspects studied in [13] or this paper. None of the participants had previous interaction with robots and none of them had previous exposure to the robot used in the experiment (NAO).

The robot used in this paper was NAO V3.3 [4] which is a small humanoid robot (Height = 57.3 cm, Width = 27.5 cm) produced by Alderbaran Robotics. Only four of the seven DoFs of each arm were controlled in this study (2DoFs in the shoulder and 2DoFs in the elbow). The lower body of the robot was fixed in a stable pose. Participant motion was collected using a Kinect sensor and the data was fed to the robot software in real time.

The experimental procedure was identical to the main study in Mohammad and Nishida [13]. We provide a brief description of the procedure here for completeness. The three conditions for the interaction were NI (No Imitation), BI (Back Imitation) and MI (Mutual Imitation) that will be explained in detail shortly. Participants had two conditions either Egyptian (EGY) or Japanese (JPN).

The experiment involved interactions between the NAO robot, the participant and a physically realistic NAO simulator (called WAN throughout the study) that was projected on a standard computer screen using Choregraphe [15]. The NAO robot and the simulator were controlled using the same software developed based on the C++ NAOqi SDK which allowed us to elicit the same motions with the same speeds from the robot and the simulator.

The experiment was designed as two rounds of a game called follow–the-leader where either the NAO robot, its simulated agent, or the participant was assigned the leader's role and the other two players tried to just copy his/its arm motion as fast and as accurately as possible.

The experimental procedure consisted – after the orientation – of three sessions of this game in the three conditions to be explained soon. A pre–experimental questionnaire (PREQ) and a post–experimental questionnaire (POSTQ) were employed as well as one questionnaire after each session (Q1, Q2, Q3). See Fig. 1 for examples from these questionnaires.

Each session consisted of two rounds. In the first round, either the robot or the simulated agent was assigned the leader's role and in the second round, the participant was always the leader and was imitated only by the robot.

The first round was the manipulated part of the experiment. Three conditions were used: BI (Back Imitation) condition in which the leader was the robot. MI (Mutual Imitation) condition in which the robot was the leader as long as the participant is accurately imitating its motion but when the participant fails in this imitation, the robot imitates the participant once then reverts to become the leader again. NI (No Imitation) condition in which the simulated agent is the leader.

The participant imitated something in the first round in all conditions (even the NI condition). What is meant by *no imitation* in the NI condition is that the participant did not imitate *the robot*. The MI condition is an extreme simplification of mother–infant mutual imitation in early years of life [7]. The only difference between the NI and BI conditions was the order by which the robot and the simulated agent moved. In the NI condition, the simulated agent moved first which made it the leader and in the BI condition the robot moved first which made it the leader. The MI condition differed from the BI condition only in that the robot occasionally (when participant's imitation was far from perfect) imitated the participant. For more information on the experimental design and justification for design decisions, please refer to Mohammad and Nishida [13].

The second round was identical in all the sessions and only the robot (not the on–screen agent) copied the pose of the subject in real time using the system proposed by Mohammad and Nishida in [10] with minor modifications. For more details on the imitation engine used please refer to [10,13]. This second round was conducted for 5 min for every participant. The same algorithm was employed with the same parameters in the three sessions which means that *objectively* the imitative skill of the robot was the same in the three conditions.

Session and pre–experimental questionnaires measured 22 independent variables (shown in Table 1) on a Semantic Differential Scale (see Fig. 1 for an example) while the POSTQ questionnaire measured the preferences of the subjects on the nine independent variables corresponding to measurement of robot skill and interaction quality. As Fig. 1 shows, the participant had the choice to select no best/worst condition.

The first five items – in all questionnaires – measured robot skill (i.e. accuracy, speed, naturalness of movement, human–likeness of motion and overall performance). One item measured participant's self evaluation of his imitative skill during the first part of the session. The remaining 16 items were the same as

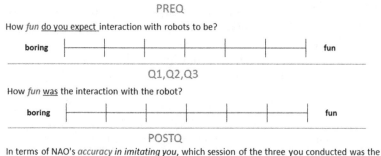

Fig. 1. Sample questions from the five questionnaires used in this study. The same questionnaires were used in [13].

Table 1. Dimensions of Evaluation employed in this study (see [13] for internal consistency evaluation).

Dimension(cronbach's α)	Dependent variable (% of variance)	Indep. variables(Loadings)
Robot skill(0.94)	Imitative skill(86)	Accuracy(0.6), overall(0.45), naturalness(0.66)
	Human–likeness of motion(6)	Human–likeness(0.995)
	Speed(5)	Speed(0.97)
Robot humanness(0.94)	Positive(70)	Curious(0.5), friendly(0.6), fun loving(0.4), sociable(0.4), *trusting(0.2)*[a]
	Negative(10)	Distractible(0.8), impatient(0.4), jealous(0.4), *nervous(0.0)*[a]
Interaction quality(0.97)	(98)[b]	Pleasant(0.7), fun(0.7)
Intention of future(0.89) interaction	(90)[b]	Closeness(0.7), living with the robot(0.7)
Likability(0.89)	Likability (84)	Polite(0.6), *sympathetic(0.02)*[a], humble(0.8)

[a]Items removed because they had small (<0.3) loadings.
[b]Dependent variable name is the same as the dimension name.

the ones used by Salem et al. to measure humanness (based on the scale designed by Haslam et al. [5]), shared-reality and likability [17].

The order of exposure to the three conditions (NI, BI, MI) was randomized between subjects. This is one difference from the study in [13] for which the higher number of participants allowed for a balancing of all ordering possibilities (12 in total). Nevertheless, the same orderings were used for Egyptian and Japanese subjects.

3   Results and Discussion

The goal of this study is to assess cultural differences between Egyptian and Japanese subjects related to back imitation. We analyzed the questionnaire data from different angles as will be shown in this section.

3.1   Effect of Imitation Condition

The first analysis step was multivariable ANOVA with nationality, imitation condition and session order as independent variables. The results are shown in Table 2. There were statistically significant effects of nationality (culture) on imitative skill (F = 5.9094(1, 46), p = 0.019) and human–likeness of motion (F = 14.5562(1, 46), p = 0.0004). There was no statistically significant effect for imitation condition which is in line with the results reported in [13] in which only preferences showed a statistically significant difference between conditions probably due to cognitive mediation.

Table 2. Multivariable ANOVA Analysis. Only dependent variables that showed statistically significant results are reported.

Dependent	Independent	F	p
Imitative skill	Condition	0.0328(2, 46)	0.9677
	Nationality	**5.9094(1, 46)**	**0.0190**
	Order	0.4694(2, 46)	0.6283
	Condition*Nationality	0.3492(2, 46)	0.7071
	Condition*order	0.9805(4, 46)	0.4275
	Nationality*order	0.4300(2, 46)	0.6531
Human–likeness of motion	Condition	1.1278(2, 46)	0.3325
	Nationality	**14.5562(1, 46)**	**0.0004**
	Order	0.9032(2, 46)	0.4123
	Condition*Nationality	**3.1434(2, 46)**	**0.0525**
	Condition*order	0.5581(4, 46)	0.6942
	Nationality*order	0.1025(2, 46)	0.9028

More interestingly for our current study there is an interaction between the experimental condition (NI, BI, MI) and nationality (F = 3.7400(2, 46), p = 0.0313) in human–likeness of motion. This interaction suggests that nationality affects the way participants' perception of human–likeness of motion was affected by the experimental condition. We found no ordering effect or interactions and this was confirmed by factorial Wilcoxon rank sum test.

We then compared the three conditions using Wilcoxon rank sum test and found no statistically significant difference between the three conditions when looking at all participants. For this test and for all factorial tests of the three experimental conditions, we use Sidak's multi comparison correction formula. Instead of the standard modification of the significance level α, we increase the individual *p-values* according to Sidak's formula assuming three tests:

$$p' = 1 - (1 - p)^3$$

This is reported as *adj.p* in all tables in this paper and is used as the basis for accepting or rejecting hypotheses. We also report the Hedge's g effect size [6] and 95 % confidence intervals for all tests.

Table 3. Results of Wilcoxon rank sum test comparing Egyptian and Japanese participants (EGY vs. JPN) based on answers to session questionnaires. Only variables that showed statistically significant results are reported.

Dimension	p	Ranksum	z	Hedges' g	95 % CI
Imitative skill	0.007	1096.50	2.68	0.743	[0.205, 1.261]
Human likeness	<0.001	1174.00	3.91	1.088	[0.526, 1.622]

To confirm the effect of nationality on subjective evaluations in session questionnaires found in the aforementioned ANOVA analysis, we conducted factorial Wilcoxon rank sum test comparing Egyptian and Japanese subjects (independent of session condition) in their evaluation of the robot. Dependent variable results are shown in Table 3 which shows statistically significant effect for imitation skill and human–likeness of motion. Both were higher for Egyptian subjects (M/SD = 4.23/1.28 and 3.53/1.26 in order) than for Japanese subjects (M/SD = 3.26/1.25 and 2.19/1.13 in order). The relation between this finding and expectations of these subjects will be discussed later in this section.

As a final check of the effect of nationality, we repeated the factorial Wilcoxon rank sum test for participants of each nationality. Egyptian and Japanese participants showed no statistically significant difference between the three experimental conditions (NI, BI, MI) in any of the measured dimensions.

Given the failure of direct participant evaluation of the robot in detecting any difference between conditions, we analyzed the preference data from participants. To analyze the preferences collected in POSTQ, we calculated a score for every session as follows (see Fig. 1 for an example question from this questionnaire): If the subject selected one session as best in some dimension, it received a +1 score. If a session was selected as worst it received a −1 score. If only a best session was selected, the remaining two sessions received a −0.5 score. If only a worst session was selected the remaining two sessions received a +0.5 score.

Finally, if no sessions were selected as best or worst (30 % of the subjects), the three sessions received a zero score.

Using Wilcoxon rank sum test to analyze this preference data, we found no statistically significant differences for the 20 participants. Analyzing the data for participants of each nationality separately showed a different story. While Egyptian participants did not show any statistically significant difference between the three conditions on any of the evaluation dimensions (Table 5), Japanese participants showed statistically significant preference for the BI and MI conditions over the NI condition for naturalness and imitative skill (Table 4).

Table 4. Results of Wilcoxon rank sum test comparing the preferences of the three conditions for Japanese participants

Dimension	Conditions	adj. p	Ranksum	Hedges' g	95 % CI
Accuracy	BI vs. NI	0.565	31.0	0.818	$[-0.490, 2.001]$
	MI vs. NI	0.165	29.0	0.935	$[-0.400, 2.126]$
	MI vs. BI	0.842	43.0	−0.196	$[-1.372, 1.010]$
Naturalness	BI vs. NI	0.088	25.50	1.685	$[0.138, 2.974]$
	MI vs. NI	**0.045**	23.50	**1.922**	**$[0.294, 3.254]$**
	MI vs. BI	0.987	41.5	−0.108	$[-1.289, 1.089]$
Human–likeness of motion	BI vs. NI	0.763	33.0	0.565	$[-0.694, 1.737]$
	MI vs. NI	0.709	32.0	0.642	$[-0.631, 1.816]$
	MI vs. BI	1.000	39.5	−0.000	$[-1.187, 1.187]$
Overall	BI vs. NI	0.444	30.0	0.980	$[-0.366, 2.174]$
	MI vs. NI	**0.022**	24.00	**2.023**	**$[0.359, 3.376]$**
	MI vs. BI	0.975	36.0	0.387	$[-0.844, 1.558]$
Speed	BI vs. NI	0.999	41.0	−0.164	$[-1.341, 1.039]$
	MI vs. NI	0.981	43.0	−0.351	$[-1.523, 0.874]$
	MI vs. BI	1.000	41.0	−0.186	$[-1.362, 1.019]$
Imitative skill	BI vs. NI	**0.045**	24.00	**1.582**	**$[0.068, 2.853]$**
	MI vs. NI	**0.026**	23.00	**2.117**	**$[0.419, 3.490]$**
	MI vs. BI	0.999	40.0	−0.049	$[-1.232, 1.143]$

For Japanese participants, Table 4 shows a statistically significant difference in naturalness (a component of the human–likeness of motion independent variable according to Table 1) between the MI and NI conditions (adj. p = 0.045) and insignificant difference with adj. p = 0.088 for BI and NI conditions. Nevertheless, only statistically significant differences in imitative skill were found in this study in preferences data. Mohammad and Nishida [13] reported – on the other hand – statistically significant differences between the same conditions for both imitative skill and human–likeness of motion. The inability to reproduce

Table 5. Results of Wilcoxon rank sum test comparing the preferences of the three conditions for Egyptian participants

Dimension	Conditions	adj. p	Ranksum	Hedges' g	95 % CI
Accuracy	BI vs. NI	0.724	22.5	0.679	[−0.751, 1.977]
	MI vs. NI	0.957	24.5	0.393	[−0.978, 1.688]
	MI vs. BI	0.930	30.5	−0.305	[−1.603, 1.051]
Naturalness	BI vs. NI	0.611	31.0	−1.095	[−2.511, 0.565]
	MI vs. NI	0.939	20.0	0.310	[−1.296, 1.835]
	MI vs. BI	0.636	12.5	1.011	[−0.941, 2.643]
Human–likeness of motion	BI vs. NI	0.977	25.5	0.333	[−1.028, 1.629]
	MI vs. NI	0.951	30.0	−0.305	[−1.603, 1.051]
	MI vs. BI	0.636	33.5	−0.808	[−2.113, 0.653]
Overall	BI vs. NI	0.092	38.00	−1.979	[−3.460, −0.115]
	MI vs. NI	0.648	33.5	−0.857	[−2.165, 0.617]
	MI vs. BI	0.611	21.5	0.748	[−0.698, 2.049]
Speed	BI vs. NI	0.370	19.5	1.150	[−0.409, 2.485]
	MI vs. NI	0.993	27.0	−0.409	[−1.792, 1.065]
	MI vs. BI	0.136	34.00	−2.097	[−3.701, −0.027]
Imitative skill	BI vs. NI	0.763	29.0	−0.496	[−1.878, 0.997]
	MI vs. NI	1.000	22.5	−0.056	[−1.603, 1.505]
	MI vs. BI	0.716	13.0	0.463	[−1.293, 2.073]

the difference in human–likeness of motion may be due to the small sample size in this study.

The conclusion we can draw from these results is that Egyptian and Japanese participants in this study differed in their response to the BI and MI conditions in comparison with the NI condition. While back and mutual imitation was associated with an increase in the perception of robot's imitative skill for Japanese subjects (as reported previously in [13]), no such differences were found for Egyptian subjects.

3.2   Analysis of Expectations and Post-Experimental Questionnaires

Table 3 shows statistically significant differences in subjective evaluations of human–likeness of motion and imitative skill between Egyptian and Japanese participants (independent of the experimental condition). This hints at a general difference in the perception of these variables by participants of the two nationalities.

To check this possibility we compared the post–experimental questionnaires and expectations of the participants of each country. Table 6 shows the results of

Table 6. Results of Wilcoxon rank sum test comparing the expectations of Egyptian and Japanese participants

Dimension	p	Ranksum	z	Hedges' g	95 % CI
Imitative skill	0.520	96.0	−0.6	0.323	[−0.597, 1.216]
Human–likeness of motion	**0.016**	74.00	−2.40	**1.090**	[0.073, 2.016]
Speed	0.847	102.0	−0.2	0.062	[−0.840, 0.960]
Interaction quality	**0.012**	72.00	−2.50	**1.309**	[0.252, 2.255]
Intention of future interaction	**0.002**	145.50	3.08	**−2.101**	[−3.155, −0.869]
Humanness (Positive)	**0.002**	146.00	3.06	**−1.923**	[−2.948, −0.734]
Humanness (Negative)	**0.007**	141.00	2.69	**−1.426**	[−2.385, −0.347]
Likability	0.306	119.0	1.0	**−0.610**	[−1.507, 0.339]

this analysis for expectations measured during the pre–experimental questionnaire using Wilcoxon's rank sum test. Egyptian subjects showed higher expectations of human–likeness of motion (p = 0.016), and interaction quality (p = 0.012) and lower expectation of both positive humanness and negative humanness (p = 0.002 and p = 0.007 respectively) and intention of future interaction (p = 0.002) compared with Japanese subjects. There was no difference in the expectation of robot speed, imitative skill or likability.

Analysis of the post–experimental questionnaire shows a slightly different pattern. Egyptian subjects now differed from Japanese subjects in imitative skill (p = 0.025), positive and negative humanness (p < 0.001 and p = 0.031 respectively). These results are summarized in Table 7. The main difference between these results and subject expectations (in the pre–experimental questionnaire) is the disappearance of the difference in human–likeness of motion, interaction quality, and imitation skill (see Table 6 compared to Table 7)

Comparing the post–experimental and pre–experimental questionnaires directly using paired t-test for both nationalities revealed no differences for Japanese subjects but a statistically significant reduction of the subjective

Table 7. Results of Wilcoxon rank sum test comparing the posterior evaluation of Egyptian and Japanese participants

Dimension	p	Ranksum	z	Hedges' g	95 % CI
Imitative Skill	**0.025**	135.00	2.23	**−1.138**	[−2.068, −0.113]
Human–likeness of motion	0.416	94.0	−0.8	0.298	[−0.620, 1.191]
Speed	0.190	122.5	1.3	−0.561	[−1.457, 0.382]
Interaction quality	0.125	125.5	1.5	−0.642	[−1.541, 0.310]
Intention of future interaction	0.878	107.5	0.2	−0.039	[−0.937, 0.863]
Humanness (Positive)	**<0.001**	153.00	3.59	**−2.768**	[−3.943, −1.358]
Humanness (Negative)	**0.031**	134.00	2.16	**−1.233**	[−2.171, −0.191]
Likability	**0.001**	149.50	3.33	**−2.091**	[−3.143, −0.862]

perception of robot's imitative skill (p = 0.008) and interaction quality (p = 0.006) for Egyptian subjects.

These results taken together suggest that Egyptian participants had higher expectations for robot's human–likeness of motion and interaction quality compared with Japanese participants, yet this difference disappeared after the experiment. On the other hand, there was a consistent difference in attribution of humanness (both positive and negative). Egyptian subjects in general attributed less humanness to the robot than Japanese subjects and this was not affected by the experiment (i.e. it appeared in both the pre–experimental and post–experimental questionnaires).

4   Discussion and Limitations

The statistical analysis reported in Sect. 3 revealed several differences between Egyptian and Japanese participants in this experiment that we will try to discuss and understand in light of known cultural differences between the two countries.

Considering expectations, Egyptian subjects showed higher initial expectation of robot's human–likeness of motion and interaction quality. This was not caused by a difference in previous experience either with the NAO itself or with other robots as all our participants reported never to have interacted with any robots before. Expectations of speed (another measure of skillfulness) did not show any difference between the two groups. Future investigation may be necessary to find an explanation for this difference or role it out on a larger sample.

A more interesting finding is that Egyptian subjects gave the robot consistently lower scores in humanness compared with Japanese subjects before and after the experiment even though they changed their scores in other factors (e.g. imitative skill and interaction quality).

Two cultural factors may be related to this difference. Firstly, the common belief that Japanese people have a more positive attitude toward robots [1] is sometimes attributed to being more secure regarding the challenge to human specialty posed by robots [8]. The common explanation is that the Shinto belief in *kami* with their mobility and existence in nature attributes to the inanimate spiritual features reserved usually for humans and the all powerful God in religions of the west (Judaism, Christianity, and Islam). This same factor may explain our finding that Japanese subjects were willing to give high humanness scores to the robot before and after interacting with it. Bartneck [1] also found that Japanese subjects prefer conventional robots over androids and the NAO robot used in this experiment was more of the first kind.

Another cultural aspect related to this finding is the traditional Islamic negative view of human–like (or even animal-like) pictures and sculptures. Most scholars of Islam ban images (*sowrah* in Arabic) depicting animate beings (e.g. humans and animals) from ritual places and many Muslim Egyptian families do not hang such pictures in their homes until today. The main justification of this negative image is that the artist when depicting animate objects is taking the role of the creator and this is considered a grave sin in Islam. This clear distinction between God-created animate beings and Human-created artificial beings may have contributed to the low expectations of humanness assigned by Egyptian participants (all Muslims) to the robot even though they had higher expectations of human–likeness of *motion* because human–like motion is not at odds with the traditional ban on human–like appearance in the artificial.

Another important finding of this paper is the interaction between culture and evaluation of the three conditions in the experiment. Egyptian participants were not affected by back or mutual imitation while Japanese subjects showed higher perception of the imitative skill and human–likeness of motion for the robot they imitated confirming the results reported in [12,13].

Two possible psychological factors were given in [13] as contributers to the effect of back imitation on perception of imitative skill and human–likeness of motion: effort justification and increased perception of agency. Effort justification refers to people's tendency to attribute greater value or importance to whatever they invest effort in and is stemming from Festinger's theory of cognitive dissonance. Effort justification should be the same for both Japanese and Egyptian participants in this experiment. This implies that the increased perception of agency may have been the critical factor. Again, culture may have something to do with this difference. Japanese people are more accepting of assigning higher humanness scores probably because of accepting a fuzzier boundary between the natural/animate and artificial/inanimate as discussed earlier. Egyptian people are less generous here probably because of the stricter boundary between the natural/animate and artificial/inanimate. Based on that, Egyptian subjects may have been less inclined to assign higher agency to the robot they imitated. It is still an artificial creation which should not and cannot have agency in this strict boundary view.

Another interesting finding of the paper is that even though Egyptian subjects had higher expectations of interaction quality and human–likeness of motion before the experiment, the short interaction with the robot reduced these unrealistic expectations causing no difference between Egyptian and Japanese participants' evaluations of these two facets of the robot after the experiment. This shows that while some cultural differences may need to just be taken into account when designing HRI scenarios (e.g. humanness assignment discussed earlier), other differences may be reduced by giving people of different cultures enough time to interact with the robot.

The findings of this paper and the previous discussion should not be taken at face value because of the limitations of this study. Firstly, the number of participants was too small to draw conclusions on the two cultures at large. Moreover, the sample was not representative of the two cultures: all participants were university students and most of them were graduate students. Moreover, all participants were students in Kyoto University which is one of the top universities in Japan. This means that they are not even fully representative of university students in these countries. This is especially true for Egyptian participants who were all, except one, graduate students living outside their country.

Nevertheless, and despite these limitations, the reported experiment hints at cultural differences between Japanese and Egyptian people in interacting with robots. The importance of these findings is enhanced by the fact that they were based on actual interaction with an autonomous robot instead of being based on pictures, videos or interactions with a WOZ (Wizard of Oz) operated robot. It enforces the message from several other studies in the HRI community that cultural aspects must be taken into account when designing robots but extends that concept to the design of the interaction instead of only appearance and robot behavior. For example, while a short session of back-imitation may be a good idea for familiarizing Japanese subjects with a robot (as we suggested in [13]), this same manipulation is not expected (based on the results of this study) to have any effect on Egyptian subjects.

5   Conclusion

In this study, expectations and effects of interaction with a NAO robot between Egyptian and Japanese subjects in a follow–the–leader game were compared. A statistically significant interaction between the participant's culture and the effect of back and mutual imitation on them was found. While Japanese subjects tended to assign higher scores of imitative skill and human–likeness of motion to the robots they imitated themselves, no such effect was found for Egyptian subjects. Moreover, analysis of the expectations and post–experimental questionnaires of the two groups revealed patterns of difference that are interesting for a follow-up study. For example, Egyptian subjects consistently assigned lower humanness scores to the robot compared with Japanese subjects. This difference may have been affected by the specific religious views common in these two cultures. In the future we will consider an expanded questionnaire that measures religious leanings of participants to provide more insight into this possibility.

Acknowledgments. This study has been partially supported by JSPS Grant-in-Aid for JSPS Postdoctoral Fellows P12046, JSPS KAKENHI Grant Number 24240023 and 15K12098, the Center of Innovation Program from JST, and AFOSR/AOARD Grant No. FA2386-14-1-0005

References

1. Bartneck, C.: Who like androids more: Japanese or us americans? In: The 17th IEEE International Symposium on Robot and Human Interactive Communication, RO-MAN 2008, pp. 553–557. IEEE (2008)
2. Bartneck, C., Nomura, T., Kanda, T., Suzuki, T., Kato, K.: Cultural differences in attitudes towards robots. In: Symposium on Robot Companions (SSAISB 2005 Convention), pp. 1–4 (2005)
3. Chartrand, T.L., Bargh, J.A.: The chameleon effect: the perception-behavior link and social interaction. J. Pers. Soc. Psychol. **76**(6), 893–902 (1999)

4. Gouaillier, D., Hugel, V., Blazevic, P., Kilner, C., Monceaux, J., Lafourcade, P., Marnier, B., Serre, J., Maisonnier, B.: Mechatronic design of NAO humanoid. In: IEEE International Conference on Robotics and Automation, ICRA 2009, pp. 769–774, May 2009

5. Haslam, N., Loughnan, S., Kashima, Y., Bain, P.: Attributing and denying humanness to others. Eur. Rev. Soc. Psychol. **19**(1), 55–85 (2008)

6. Hedges, L.V.: Distribution theory for Glass's estimator of effect size and related estimators. J. Educ. Behav. Stat. **6**(2), 107–128 (1981)

7. Jones, S.S.: The development of imitation in infancy. Philos. Trans. R. Soc. B: Biol. Sci. **364**(1528), 2325–2335 (2009)

8. Kaplan, F.: Who is afraid of the humanoid? investigating cultural differences in the acceptance of robots. Int. J. Humanoid Rob. **1**(03), 465–480 (2004)

9. Lee, H.R., Sabanović, S.: Culturally variable preferences for robot design and use in south korea, turkey, and the united states. In: Proceedings of the ACM/IEEE International Conference on Human-Robot Interaction, HRI 2014, pp. 17–24. ACM (2014)

10. Mohammad, Y., Nishida, T.: Tackling the correspondence problem. In: Yoshida, T., Kou, G., Skowron, A., Cao, J., Hacid, H., Zhong, N. (eds.) AMT 2013. LNCS, vol. 8210, pp. 84–95. Springer, Heidelberg (2013)

11. Mohammad, Y., Nishida, T.: Effect of back and mutual imitation on human's perception of a humanoid's imitative skill. In: The 23rd IEEE International Symposium on Robot and Human Interactive Communication, RO-MAN 2014, pp. 788–795. IEEE (2014)

12. Mohammad, Y., Nishida, T.: Human-like motion of a humanoid in a shadowing task. In: International Conference on Collaboration Technologies and Systems, CTS 2014, pp. 123–130. IEEE (2014)

13. Mohammad, Y., Nishida, T.: Why should we imitate robots? effect of back imitation on judgment of imitative skill. Int. J. Social Robot. **7**(4), 497–512 (2015). http://dx.doi.org/10.1007/s12369-015-0282-2

14. Nielsen, M.: Imitation, pretend play, and childhood: essential elements in the evolution of human culture? J. Comp. Psychol. **126**(2), 170–178 (2012)

15. Pot, E., Monceaux, J., Gelin, R., Maisonnier, B.: Choregraphe: a graphical tool for humanoid robot programming. In: The 18th IEEE International Symposium on Robot and Human Interactive Communication, RO-MAN 2009, pp. 46–51. IEEE (2009)

16. Riek, L.D., Paul, P.C., Robinson, P.: When my robot smiles at me: enabling human-robot rapport via real-time head gesture mimicry. J. Multimodal User Interfaces **3**(1–2), 99–108 (2010)

17. Salem, M., Eyssel, F., Rohlfing, K., Kopp, S., Joublin, F.: To err is human (-like): effects of robot gesture on perceived anthropomorphism and likability. Int. J. Soc. Rob **5**(3), 1–11 (2013)

18. Salem, M., Ziadee, M., Sakr, M.: Marhaba, how may i help you?: effects of politeness and culture on robot acceptance and anthropomorphization. In: Proceedings of the ACM/IEEE International Conference on Human-Robot Interaction, HRI 2014, pp. 74–81. ACM (2014)

19. Samani, H., Saadatian, E., Pang, N., Polydorou, D.: Newton Fernando, O.N., Nakatsu, R., Valino Koh, J.T.K.: Cultural robotics: The culture of robotics and robotics in culture. Int. J. Adv. Robot. Syst. **10**, 400–409 (2013)

20. Shibata, T., Wada, K., Tanie, K.: Statistical analysis and comparison of questionnaire results of subjective evaluations of seal robot in Japan and UK. In: IEEE

International Conference on Robotics and Automation, ICRA 2003, vol. 3, pp. 3152–3157. IEEE (2003)

21. Taras, V., Rowney, J., Steel, P.: Half a century of measuring culture: review of approaches, challenges, and limitations based on the analysis of 121 instruments for quantifying culture. J. Int. Manag. **15**(4), 357–373 (2009)
22. Trovato, G., Zecca, M., Sessa, S., Jamone, L., Ham, J., Hashimoto, K., Takanishi, A.: Cross-cultural study on human-robot greeting interaction: acceptance and discomfort by egyptians and japanese. Paladyn, J. Behav. Rob. **4**(2), 83–93 (2013)
23. Wang, L., Rau, P.L.P., Evers, V., Robinson, B.K., Hinds, P.: When in rome: the role of culture & context in adherence to robot recommendations. In: ACM/IEEE International Conference on Human-Robot Interaction, HRI 2010, pp. 359–366. IEEE Press (2010)

Designing the Appearance of a Telepresence Robot, M4K: A Case Study

Hyelip Lee[1]([⊠]), Yeon-Ho Kim[2], Kwang-ku Lee[2], Dae-Keun Yoon[2], and Bum-Jae You[2]

[1] Department of Industrial Design, KAIST, Daejeon, South Korea
lhlilalee@kaist.ac.kr
[2] Center of Human-centered Interaction for Coexistence, KIST, Seoul, South Korea

Abstract. This paper presents the process for designing a telepresence robot (TPR) – M4K (Mobile 4-Dimensional Communication Kiosk), which offers more extensive interaction between users than other existing TPRs. The TPR is a robotic movable platform which helps people communicate across distances, and it is sometimes required to have the ability of 'tele-manipulation', which exceeds the current capacity of pure 'telecommunication' [13]. Of note, our M4K has an interactive 3D beam projector and cameras in addition to the basic TPR platform. The design process to give the M4K an acceptable form to users is presented. Following illustrating the design process, limitations and future works are discussed. The design process progresses from related research, ideal version design, to manufactured version design. This project is ongoing and at an estimated midway point of the whole development cycle, so it is also treated partly as an experiment. This study will be used to support the next phase of design and technical development.

1 Introduction

The concept of a 'telepresence system' is to convey a realistic sense of 'being there' to users who access the system from remote locations, and who interact with it. Telepresence robots (TPRs), also called mobile robotic telepresence (MRP) systems, are characterized by a video conferencing system mounted on a mobile robotic base [6]. The TPR consists of the physical robot and the interface used to control the robot. It helps an operator feel like he/she moves around in the robot's environment even from a faraway place. It also provides the user with a sense of social interaction with a person in another location.

Forming part of the projects conducted at the Center of Human-centered Interaction for Coexistence (CHIC), M4K (Mobile 4D+ Communication Kiosk) shown in Fig. 1 has been developed as an upgraded version of the mobile robot-based 3D interaction platform [14]. It shares similar equipment to ordinary TPRs, such as a monitor, camera, and mobile platform. In addition, it contains a beam projector which can display 3D images to wherever users direct it (floor, wall, ceiling, etc.). These images can be manipulated through an interactive glove or pen. Maintaining these functions, we tried to work on the appearance

© Springer International Publishing Switzerland 2016
J.T.K.V. Koh et al. (Eds.): Cultural Robotics 2015, LNAI 9549, pp. 33–43, 2016.
DOI: 10.1007/978-3-319-42945-8_3

Fig. 1. M4K

of the equipment to make it appear more accessible to the general public. To make a persuasive design even to researchers and technicians, it was necessary to survey and study an ideal version of a robot form. The ideal version was then transformed into an actual version, with fixed equipment and an undetermined context (to make it ubiquitously accepted). The design schedule was initially estimated at three weeks, but became extended to more than a month. In this paper, the M4K's design background, process, result, and future works are discussed.

2   Related Work

2.1   Robot Appearance Design

Recent research papers about TPRs [3,6] were mostly focused on the technical aspects such as movement speed, communication stability, adjustable autonomy, and also psychological aspects such as social and emotional interactions [7,12], while one of the core factors, appearance, was rarely mentioned. Consideration of the physical appearance appeared in some papers [8,9], but they tended to cover details about interaction methods rather than the exact appearance of the design. The papers related to appearance are discussed here.

Social Robots Design. When designing a social robot, Fong et al. [4] considered that the physical appearance of a robot is important and that it should be matched with its intended function because it can bias the interaction with users' expectations towards the robot. Not only did they mention Mori's "uncanny valley", which showed the robot's proper similarity to humanness that affects familiarity, they also classified social robots according to four broad categories:

Anthropomorphic, Zoomorphic, Caricatured, and Functional. For example, they asserted that a caricatured design can make a robot be more focused on the desired interaction, making certain features strong or weak. They also presented other researchers' arguments in that there are robots such as health care robots that should reflect their function. From such literature, it was decided that M4Ks character should have either a caricatured or functional design, because M4K has a rather utility-like character with the main function of telecommunication rather than being an independent, individual character.

Robot Form and Operator's Self-Extension. Groom, et al. [5] compared the impact of robot form (humanoid robot vs. car robot) on users' feelings of self-extension toward the robots. Participants of the experiment showed greater extension of their self-concept in the car robot, and they preferred the car robot more than the humanoid robot. The reason given was that people perceived the humanoid form as an entity with its own identity. The researchers suggested the implication of the experiment results on certain situations.

In the case of a telepresence robot, users feel a non-humanoid form will be better when it represents the person who connects to the system, for example, when doctors examine patient remotely through the robot. In that case, promoting self-extension will be more useful to improve the interaction between users. On the other hand, if the telepresence robot is used for a dangerous situation such as a rescue or hostage negotiation, a humanoid form will be more appropriate in that it minimizes self-extension and protects the operator's state of mind. This strengthened the belief to make the M4K more product-like or character-like in form as opposed to a humanoid form. The M4K is designed for extending the operator's feeling of self-existence at other locations, and this is more appropriate for connecting people in everyday social contexts.

Robot's Height. Referring to the impact of height on the interaction between humans, Rae et al. [10] investigated the effect of the telepresence robot's height on the local user's perceptions of the operator and subsequent interactions. The importance of understanding the potential users of the system and their roles before designing the TPR was highlighted. For example, to make the robot appear more agreeable and cooperative, a shorter height than the local user is needed. In contrast, a taller height is needed to convey the sense of an important, specialist role to the robot. From this study, it was felt flexible height will be appropriate because the role of TPR operator can then be changed according to circumstance. If the height had to be fixed, however, we decided to set it around 1200 mm in order to be similar to a seated person.

2.2   Current Telepresence Robot Design

Through understanding the TPRs hitherto developed, the ideal characteristics of TPRs could be considered before designing the M4K. Kristoffersson et al. [6], mainly talk about MRP systems which have non-anthropomorphic appearance with a limited subset of human skills (e.g., pan-tilt, simple gesticulation). The heights in those systems are mostly similar to or smaller than a normal adult's height. Kristoffersson et al. also show anthropomorphic designs. Comparing the two different approaches, it was felt that a mid-point between the two should be considered in order to make a more familiar form while retaining its image of a useful tool at the same time. Materials such as fabric and silicon were discussed, but owing to time, at this stage it was mocked up with standard plastic. Material research will take more time to better understand its capability.

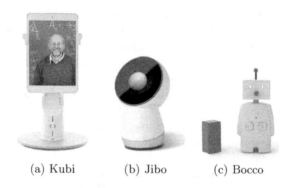

(a) Kubi   (b) Jibo   (c) Bocco

Fig. 2. TPRs with small shape, non-mobile function

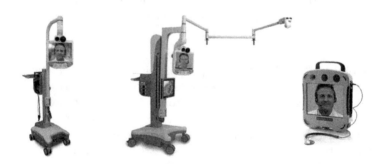

Fig. 3. TPRs with medical functions (InTouch Lite, Vantage, Xpress)

In addition, according to Tsui et al. [13], TPRs have a lighter weight and cost less when designed for general usage. The structure of the head monitor part is sometimes designed to be removable and sometimes incorporates a tablet device.

Figure 2 shows TPRs which have a small shape, such as those that can be put on a table. They do not have a mobile function. They have lighter or more familiar shapes than previous TPRs. Among them, Jibo [2] has a telecommunication

Version 1.0   Version 2.0

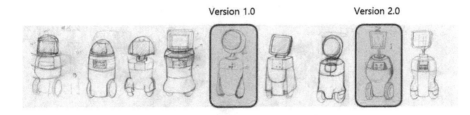

Fig. 4. Ideal version's early sketches

function among other functions. Also, the TPRs from InTouch Health [1] are shown in Fig. 3. Those were developed for specialized medical environments. If the later M4K's movement range is not so wide or its usage is specific to medical environments, such size, weight, and shapes will be considered.

3   Design Process

Firstly, an ideal version of a robot form was suggested. Then the design was matched with currently developed equipment and budget constraints. This process is not in fact atypical to any other product development process. However, it is important to regularly review the development process, as the design still has some way to go before completion.

3.1   Ideal Version

By applying only the character and composition of the equipment, the ideal version had a somewhat undetermined final total design. In this stage, the size of the equipment and other specific technical parts were ignored.

To understand the given information about M4K, we reorganized the information according to the seven design dimensions suggested by Rae et al. [11] as shown in Table 1. Reflecting the current research mentioned above, one goal of the design was to have a caricatured or functional form, more of a mid-point shape between non-anthropomorphic and anthropomorphic, and to have a height similar to a seated person's height. Maintaining the upper monitor's location, it was decided that the monitor should be embedded and not detachable. This ideal version played a useful role in helping communication with technicians afterward.

Through the sketches, as shown in Fig. 4, the design direction was discussed. The specific design was developed after choosing one of the options. The ideal version shown in Fig. 5 is named *Aesop*. The form contains curved shapes to offer a familiar humanistic feeling. LED lighting to express the robot's state is located along the side of a head screen, and the top of the head. A speaker is embedded on the underside of the neck. The body part has round shape that reduces edge dislocation, and the wheel part is minimally exposed in order to hide the mechanical properties.

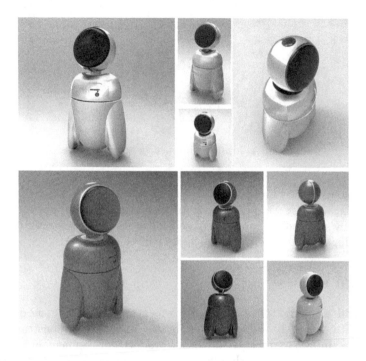

Fig. 5. Ideal version of M4K

Evaluation. For enabling stable movement, a stanchion or one more wheel in the middle is needed. Since the design itself may be close to other existing robots, a more distinctive design is required. There are technologically restricted parts such as controlling the location of the projected image through only the movement of the projector's lens.

3.2   Process and Design Changes

After designing the ideal version, a realization process was conducted reflecting some of the limitations. Table 2 shows the process.

Form. For offering a familiar feeling, a curved form was suggested, but it changed to a more neutral and angular form in order to have more continuity with the lower part. The lower part's equipment was uncovered to maintain its size.

Color. From the initial home and family concept of usage for M4K, it became extended to more general contexts including offices and schools. Accordingly, the color of the robot changed from warm and bright colors to a metallic and calm tone while considering it as a social communication tool. With those considerations, several color samples were suggested as shown in Fig. 6.

Table 1. Characteristics of M4K in seven design dimensions

Seven design dimensions	M4K
Initiation	Unknown
Physical environment	Indoor
	Home + Conference room
Mobility	Non-holonomic
Vision	Webcam, screen view
Social environment	Family members, coworkers
Communication	Fun, work
Independence	Unknown
Other	Beam projector (3d)
	4Degree of freedom (base, neck, head, center parts)

Fig. 6. Several color samples

3.3 Final Version

We selected the two final colors from a survey conducted at the research center, and the last one, Fig. 7 was chosen and manufactured. Figure 8 shows the mock-up design pieces and Fig. 9 is the final manufactured view of M4K. The color became further simplified.

After making the final version (Fig. 9), the design was critiqued. The color appears well-matched with the purpose, and the covers do not disrupt the movement. It has the caricatured shape, and it is more product-like than human-like as was intended. However, the size and proportion of the head should be reduced. Redesign of the equipment parts is needed to make a slimmer shape. Assembly methods also need to be simplified.

3.4 Contribution and Limitation

This paper focused on a TPR's appearance design, which was rarely discussed in previous research. It started with Human robot interaction (HRI) research, and

Table 2. Change of M4K's form–timeline

Date	Description
Mar. 02, 2015	The lower part's equipment already has a large size. Therefore, we decided to expose the lower part and to cover the upper part. Firstly, only using the equipment's images, we explored a simple and feasible form. While maintaining the ideal version's circular face, we designed the inner body cover separately, considering the beam projector's up-down movement. *Assumption*: Its main user is children and adults (home environment)
Mar. 06, 2015	 Now designed, with the specific size of the equipment in place. We decided to paint the lower part to be harmonized with the upper part. The monitor part became more square-shaped to cover the existing monitor and the body's inner cover developed to have a caricatured form. The caricatured facial image moves from monitor to inner cover as version 2.0 in Fig. 5.
Mar. 09, 2015	Then showing the covered upper part and painted lower part together. Considering consistency with the lower equipment part's character, the simple figure-shaped upper part was suggested. An effort was made to minimize the monitor's cover size.
Mar. 13, 2015	Considering the blind spot, the edge of the monitor screen was uncovered. The monitor cover's shape followed the body shape, so it has a simple figure shape. To reduce the size of the body part, the backside was cut out.
Mar. 19, 2015	Color changes are made.

this showed the value of the ideal version design, which worked as a motivator for discussion among researchers. The rest of the design process details how the final version was manufactured.

There was a difficulty in designing robot appearance with an almost fixed hardware platform without understanding specific users' needs. There was a basic design purpose - TPR - but a more detailed scenario would be useful. The extended concept of general usage was used at the coloring stage. It would have been desirable to have considered this at an earlier stage. We went directly from 3D computer modeling to a real size mock-up, but ideally, if there was more time,

Fig. 7. Final version 3D rendering

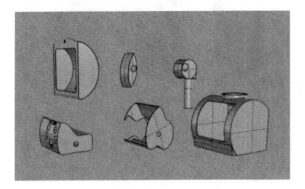

Fig. 8. Mock-up pieces

a small-scaled physical model would have been useful to capture the feelings of the final form before making the actual-sized mock up.

4   Future Work

When existing TPRs are mainly used in offices, hospitals, and schools [6], greater understanding about the real demands and situations of these environments are required to design properly. If we understand the users' needs correctly, it will be possible to consider not only the robot's appearance, but also to consider finer interactions between a human and a robot. That interaction will also change the robot's shape. Accordingly, there is a plan to conduct necessary surveys (e.g., interview, questions) at each of the possible user environments. Then, comparing the ratio of the common needs against special needs, it will be possible to judge the design direction of the TPR for general or special usage. Now that the M4K project is in the middle stage of the whole development cycle, Material and animation research should go together.

Fig. 9. Manufactured version

5 Conclusion

This paper demonstrated TPR design: the development process behind M4K's appearance, with a discussion of the results, difficulties, and future works. This paper will be an important reference to the final design and technological development. Furthermore, not only for this project, but also for other TPR or general robot designs, this paper can be used partially and critically. When most existing robots are often down to an individual designer's sense, our experience of design process demonstrates how it has been useful to consider different factors and existing literature when designing a product that will be better liked by users.

References

1. Intouch health. http://www.intouchhealth.com
2. Jibo. https://www.jibo.com
3. Desai, M., Tsui, K.M., Yanco, H., Uhlik, C., et al.: Essential features of telepresence robots. In: 2011 IEEE Conference on Technologies for Practical Robot Applications (TePRA), pp. 15–20. IEEE (2011)
4. Fong, T., Nourbakhsh, I., Dautenhahn, K.: A survey of socially interactive robots. Robot. Auton. Syst. **42**(3), 143–166 (2003)

5. Groom, V., Takayama, L., Ochi, P., Nass, C.: I am my robot: the impact of robot-building and robot form on operators. In: Proceedings of the 4th ACM/IEEE International Conference on Human Robot Interaction, pp. 31–36. ACM (2009)
6. Kristoffersson, A., Coradeschi, S., Loutfi, A.: A review of mobile robotic telepresence. Adv. Hum. Comput. Interact. **2013**, 3 (2013)
7. Lee, M.K., Takayama, L.: Now, i have a body: uses and social norms for mobile remote presence in the workplace. In: Proceedings of the SIGCHI Conference on Human Factors in Computing Systems, pp. 33–42. ACM (2011)
8. Leithinger, D., Follmer, S., Olwal, A., Ishii, H.: Physical telepresence: shape capture and display for embodied, computer-mediated remote collaboration. In: Proceedings of the 27th Annual ACM Symposium on User Interface Software and Technology, pp. 461–470. ACM (2014)
9. Rae, I., Takayama, L., Mutlu, B.: In-body experiences: embodiment, control, and trust in robot-mediated communication. In: Proceedings of the SIGCHI Conference on Human Factors in Computing Systems, pp. 1921–1930. ACM (2013)
10. Rae, I., Takayama, L., Mutlu, B.: The influence of height in robot-mediated communication. In: Proceedings of the 8th ACM/IEEE International Conference on Human-Robot Interaction, pp. 1–8. IEEE Press (2013)
11. Rae, I., Venolia, G., Tang, J.C., Molnar, D.: A framework for understanding and designing telepresence. In: Proceedings of the 18th ACM Conference on Computer Supported Cooperative Work & Social Computing, pp. 1552–1566. ACM (2015)
12. Tsetserukou, D., Neviarouskaya, A.: Emotion telepresence: emotion augmentation through affective haptics and visual stimuli. J. Phys. Conf. Ser. **352**, 012045 (2012). IOP Publishing
13. Tsui, K.M., Yanco, H.A.: Design challenges, guidelines for social interaction using mobile telepresence robots. Rev. Hum. Factors Ergon. **9**(1), 227–301 (2013)
14. You, B.-J., Kwon, J.R., Nam, S.-H., Lee, J.-J., Lee, K.-K., Yeom, K.: Coexistent space: toward seamless integration of real, virtual, and remote worlds for 4d+ interpersonal interaction and collaboration. In: SIGGRAPH Asia 2014 Autonomous Virtual Humans and Social Robot for Telepresence, p. 1. ACM (2014)

The Positive Effect of Negative Feedback in HRI Using a Facial Expression Robot

Mauricio Reyes[1]([⊠]), Ivan Meza[2], and Luis A. Pineda[2]

[1] Centro de Investigaciones de Diseño Industrial (CIDI), Universidad Nacional Autónoma de Mexico (UNAM), Mexico City, Mexico
mauricio@turing.iimas.unam.mx
[2] Instituto de Investigaciones en Matematicas Aplicadas y en Sistemas (IIMAS), Universidad Nacional Autónoma de Mexico (UNAM), Mexico City, Mexico
ivanvladimir@turing.iimas.unam.mx, lpineda@unam.mx

Abstract. This research explores the use of facial expressions in robots and their effect in collaborative tasks between humans and robots. The positive effect is determined during a task in human - robot collaboration, derived from a negative facial expression issued as feedback by the robot (sad face) when a failure in the execution of the task occurs. This study analyzes whether or not human intervention exists on the initial presence of an unexpected failure, the response time of the intervention and the accuracy of the task. A comparison with a neutral facial expression is also performed.

Keywords: Collaborative interaction · Minimalist robot head · Negative facial expression · Reaction time

1 Introduction

Nonverbal social cues help strengthen the interaction between humans and robots. Among the most explored ones are: pointing, observing, body language and facial expression. The role of emotions in human behavior has led to develop emotional computational analogues, capable of driving more intelligent and flexible systems in complex and uncertain environments [1]. Emotions can be identified as the means that serve individuals to establish values and improve their interaction and outcome. Based on emotional displays, individuals who perceive the emotions can create interpretations or assume the state of the person who shows them [2].

The human face serves many purposes, it shows an individual's motivation, which helps create a more predictable and understandable behavior for others; this is supported by sight and eye tracking. During human collaboration, gestures play a key role in communication, used daily to maintain, invite, synchronize, organize or finishing a particular activity [3]. Artificial emotions in robots and virtual agents favor feedback during interaction with humans [4]. Emotional models in robotic architectures are developed to make explicit the goals of robots and strengthen the link between person and robot [5]. This work explores the effect of a facial expression used as feedback and its positive effect on a cooperative task (human-robot interaction). Our study focuses on the

© Springer International Publishing Switzerland 2016
J.T.K.V. Koh et al. (Eds.): Cultural Robotics 2015, LNAI 9549, pp. 44–54, 2016.
DOI: 10.1007/978-3-319-42945-8_4

influence of a negative facial expression (*sad*) on reaction and accuracy of human actions. In particular, we study the first sudden failure and the primal effect on the interaction.

2 Background

The uses of social signals integrated into robotic systems have been crucial to the HRI community. These signals are useful at the coordination of actions between humans and robots and help interpreting human reasoning and cognitive patterns. The motivation that gives meaning to the correct study of such factors is to improve the collaborative activities with common goals between humans and robots. It also measures and quantifies the impact of the interactions and influence on people [6]. Various aspects such as behavior, physical appearance, natural language, gazing, pointing, gestures, personality and body language might be related to the way humans empathize and perceive robots. It is considered that the use of nonverbal signals by robots improve social interaction with people. For instance, the robot's gaze can communicate care and caution, and express with facial expressions an emotional state [7].

The use of social signals have been extensively studied through robotic faces aiming to improve collaborative work focusing on the use of the gaze to coordinate and strengthen the interaction as well as predict and clarify intentions among colleagues, [7, 8], which influence accuracy and speed of human action [9]. With regard to robotic expression of emotions and facial gestures, multimodal communication in robots have been explored to create comfortable social interactions within complex environments such as interactive museums [10] and public displays [11]. In the field of manufacturing, it has been shown that the use of gestures in industrial robots contain a wide communicative content for the human parts. The readability of the robot gestures has a positive influence on an industrialized collaborative task, reflected in the continuity of an activity, reducing work-time and increasing productivity [12]. Nonverbal social cues may be beneficial in the interaction with robots used for specific tasks oriented towards service robots [13].

Human perception of gestures influences communication and discrimination of positive and negative effects. Identification of negative facial gestures (*sad*) affects the viewer's attention more effectively than a positive gesture (*happy*), regardless of whether the gesture identification search is intentional or unintentional [14]. Using feedback from negative social effects on robotic agents and the importance of positive effects on the job or part of the structure of it has been suggested for the design of robotic agents capable of altering their behavior [15].

3 Robot Face

We developed GolemX-1, a minimalistic robotic face, capable of performing the basic emotions described by Ekman [16]. We consider the minimum of elements, fabrication accessibility and efficiency in readability of expressions for our robot's face since this will be used in the Golem II and III (Fig. 1) service robots [17]. The robotic face was inspired on some of the characteristics of the MiRAE robotic face and its minimal component development [18]. MiRAE has been tested with the Facial Expression Identification (FEI) rates for each of their expressions. MiRAE robotic face has the following features: (a) Ability of facial expression based on the use of six horizontal lines, (b) Reduction in the complexity of construction (c) Readability of interpretation by human subjects of the basic emotions described by Ekman.

Our robotic face was limited to the readability of facial expressions with elements similar to those minimum requirements of the MiRAE robot. This resulted in a robotic face in which we avoid the excessive use of aesthetic elements and degrees of freedom (DOF) used in other robotic faces [19, 20]. Our robotic face is built with electromechanical equipment, manufacturing material and accessible construction processes at low-cost, easy maintenance and the electronic control equipment Arduino Uno[1]. GolemX-1 is capable of running both negative (sadness, anger, fear) and positive expressions (happiness, surprise), using 7 DOF. Our robot's face was subjected to a multiple choice online study to test the readability of the face and adjustment execution. For this study only 3 types of facial expression were used: Neutral, happiness and sadness (Fig. 2). Additionally, an upper body with arms was built to adequate the experiment for human-robot interaction, one of the hands is movable with 1DOF.

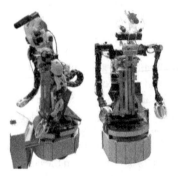

Fig. 1. Golem-III service robot.

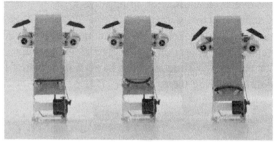

In order (Left to Right) Neutral, Happiness and Sadness expressions.

Fig. 2. GolemX-1. Expressive robotic face. In order (Left to Right) neutral, happiness and sadness expressions.

---

4 Experiments

Our experimentation assumes the following:

- The use of gestures (facial expressions) is useful in the communication between humans and robots in a collaborative task.
- Facial expressions performed by a service robot provide feedback about achieved or failed progress in tasks
- The use of gestures in a service robot has influence over the human actions in a collaborative task.

Taking in to consideration the previous aspects we evaluate the readability of the GolemX-1 expression and the effect of the negative expression (*sad*) on human behavior when the robot fails to perform a task correctly. In the case of readability we followed an online survey methodology. In the case of the measuring the effect we design a teamwork scenario between human and robot. We focused on the first failure of the robot since it captures the naive response of the human subject in a collaborative interaction.

4.1 Facial Expression and Readability Test

The objective of this experiment was to validate the readability of facial expressions issued by GolemX-1's robotic face. The following hypothesis was established:

H1: *The robotic face performs the negative facial expression of sadness in a legible manner.*

To test H1 we designed an online survey of facial expression identification. We showed to a subject a digital version of GolemX-1 expressing emotional gestures (*neutral, happy, sad, worry, surprise and fear*). We promoted the online survey among students and colleagues, 88 subjects answered the survey. We showed the robot faces with the six selected gestures randomly and one by one for the participant to choose an emotion from a list with random labels in each image. We invited several subjects from which 88 answered the survey. The majority of subjects were industrial design students; few of them were engineers in computer science. The average age was 24 years old.

4.2 Collaborative Task Experiment and Scenarios

The aim of the experiment was to determine the influence of negative facial expression of a robot on the human response, only when an unexpected failure occurs for the first time during a task. The experiment was designed to explore with non complex repetitive task the influence between the facial expression and the human actions and if this influence can be registered and significant in order to use in a collaborative human - robot task. The non complex execution of the task increases the control during an unexpected failure. Although in the experiment the robot face used facial expressions as main social cue, it also used its gaze as the robot's attention indicator, which also allows smoother interactions. However, we avoided to alternate the direction of the gaze between human and object because the gaze is used to trigger human interactions in some collaborative task under different conditions [9]. The robot´s actions time response was empirically set so was suitable for human response after executing several test sessions.

The following hypothesis was stated:

H2: A negative robotic facial expression can influence the behavior of human actions.

This experiment consists in a simple task with shared responsibility for the execution. The human and the robot cooperate to place ten cylindrical objects within a container. However we have programmed the robot to randomly fail to place the object in the container. An electromechanical system of interaction was used as workspace and as interface of the human-robot collaboration. The hand with 1DOF integrated to the upper body of the robot was also used to manipulate the objects. Figure 3 shows the areas and system elements. At the start of the experiment interaction the workspace is placed at the center of a table.

The human subject is located on one side of the table with the ten cylindrical objects. The robot, GolemX-1, is facing the human subject on the opposite side of the table. The robot's right hand is by the interaction system, beside the launching area, waiting to push the objects through the ramp that leads to the container. The subject must wait for the access barrier to open to place an object in front of the robot in the launching area. The subject is instructed to put only one object every time the barrier opens. Then, the access barrier closes and the robot pushes the object towards the ramp into the container. The subject cannot access or intervene in the action unless the barrier is open. Two video cameras were used to record the interaction. The experiment procedure is performed as follows:

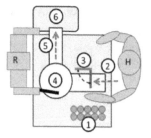

Fig. 3. Workspace scenario. (Left) Interaction system and collaborative Human-Robot workspace diagram: (1) Objects zone, (2) Human access, (3) Access barrier (closed/open), (4) Launch area and robotic hand, (5) Ramp, (6) Container. (Right) Human-Robot setup and workspace. (Color figure online)

- Written instructions are given to the user. The researcher makes a test session to exemplify the general movements. This was done by placing the object in front of the robot. Once positioned, the barrier is closed; the robot observes the object and successfully throws it into the container. The robot starts the interaction. The barrier is opened and the robot faces forward.

- The human subject takes his place and the interaction starts.
- The subject places the first object in front of the robot. In the first try the robot pushes the object in to the container and a successful outcome occurs.
- The human place the next object and a successful outcome is repeated.
- The robot actions are randomized from the third execution onwards. A failure can occur at any point of the interaction. A failure is when the robot is not able to push the object. During a failed action, the robot looks at the object but the hand movement is limited to prevent the object from being pushed into the container. The robot does not look at the human again. Duration of the failed action: 9 s.
- After the failed action, the access barrier is opened for 11.5 s in order to give an opportunity to the human user to intervene in the work of the robot. During this time the robot continues gazing at the object.
- At the end of the allocated time the barrier is closed, the failed action is repeated once again and the barrier is reopened by 11.5 s. This is done in order to motivate a reaction from the human.
- The task finishes until the ten objects fall into the container.

Two types of scenarios were mounted using this same procedure. In the first one, the robot executes a neutral expression for the task. In the second one, the robot performs a positive facial gesture (happy) when the task is successful and a negative facial gesture (sad) when the task fails. The detailed actions are specified in Table 1. The experiment was conducted with 15 naïve human subjects (9 female, 6 male) for the scenario with a neutral expression; and with 15 subjects (7 female, 8 male) for the facial expressions scenario. All subjects were industrial design students and the average age was 22 years old. They were recruited by personal invitation.

Table 1. Task description.

Actions in general task	Collaborative human-robot task		
	Human action	Robot action neutral face	Robot action expressive face
Start (Successful or Failed action)	The access barrier is open. The human subject puts the object in front of the robot	Looks at the object	Turns its head toward the object. Neutral face
Successful action	The access barrier is closed. The human subject waits	Moves its hand and pushes the object. Returns its hand to the start position	Starts the positive facial expression. Moves its hand and pushes the object. Returns its hand to the start position

(Continued)

Table 1. (*Continued*)

Actions in general task	Collaborative human-robot task		
	Human action	Robot action neutral face	Robot action expressive face
Finish successful action Restart	The access barrier is open The human subject can put other object	Looks at the human waiting for the next object	Finishes the positive facial expression. Turns its head toward the human and waits for the next object
Failed action	The access barrier is closed The human subject waits	Is not able to push the object Waits for 9 s	Is not able to push the object. Starts the negative facial expression for 9 s
Finish failed action	The access barrier is open for 11.5 s	Remains looking at the object	Finishes the negative facial expression. Remains looking at the object

As a random programmed activity the robot gets a success after 1 or 2 attempts more in order to avoid jamming the task.

5 Results

5.1 Readability Test

The confusion matrix for the readability and accuracy of the neutral, sad and happy gestures are presented in Table 2. The rest of the gestures were omitted because they were not used during the experiment but they follow a similar trend. However, the gestures have to be improved in order to measure the impact of the positive effect. These observations are confirmed once we calculate the accuracy of the gestures. From the table it can be noticed that neutral and sad gestures can transmit the intended emotion with accuracy. On the other hand, the happy gesture is less effective. We consider this is not relevant for our following experiment since we did not focus on this gesture at the moment.

5.2 Collaborative Task Experiment and Scenarios

The reached accuracy for the sad expression of 85 % allows using GolemX-1's robot face in order to study the negative feedback convey by the sad gesture in a collaborative task. In particular, this experiment focus on the first robot failure to push the object into the container. For this we focus on the behavior of the human when this happened. This situation defined a specific failure pattern which was analyzed in the following aspects:

- We identify the type of behavior during the human subject intervention to the failure (or lack of it).

- The reaction time form the opening of the barrier to the time the subject intervenes during a failure.
- The performance of the subjects following the instructions of the task.

We considered that an intervention from the subject happened if she or he tried to help during the course of the task. By analyzing the video recordings of the interactions, four types of intervention were identified in case of failure: (a) The subject changed the position of the object. (b) The subject changed the current object for another from the stash. (c) The subject placed an extra object into the launching area. (d) The subject waited. Table 3 presents the frequency these behaviors shown in the scenarios. Using a neutral facial expression feedback produces the subject to place more objects or to wait more frequently than when using a negative facial expression feedback. Both putting more objects and waiting go against the flow of the task, for instance, making sessions longer. In Table 4 we can see the frequency on terms of the behaviors being positive or negative to the continuity of the task. Both behaviors have different statistically significance when compared ($p < 0.05$)[2].

Table 2. Facial expression test and accuracy

Response	Facial expression			
	Neutral	Happy	Sad	Accuracy
Neutral	**67**	11	3	76.1 %
Happy	0	**56**	1	63.3 %
Sad	6	0	**75**	85.2 %
Others	15	21	9	–

"Others" correspond to angry, fear, surprised and worried facial expressions.

Table 3. Type of action and intervention on task

Action-intervention	Neutral face experiment	Negative face experiment
The human subject changed the object's position	7	14
The human subject changed the object	1	1
The human subject put more objects in the launch area	4	0
The human subject waited until the robot gets a success	3	0

By analyzing the reaction times from both scenarios we notice that when only using the neutral feedback, the average reaction time is longer than when using the negative

---

[2] Statistical significance calculated by Exact Fisher method for an experiment with a power of 0.9 for a sample of 15 per case.

Table 4. Positive and negative effect

Type of robotic face in experiment	Effect in the task	
	Neutral feedback	Negative feedback
Positive effect	8	15
Negative effect	7	0

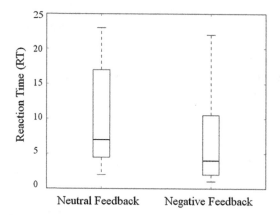

Fig. 4. Time distribution on reaction time.

feedback (11 s, neutral; 7 s, sad). However, when doing a more strict analysis we don't find evidence of statistic significance ($p > 0.05$)[3] (Fig. 4).

6   Conclusions

This research explores the effect of negative facial expression feedback during a collaborative human-robot interaction task.

First, we measured the readability of the sad facial expression of GolemX-1 robotic face using an online survey. We founded that this expression was one of the best expressed gestures. Since it has been shown that this gestures influences the attention of the human subject [21] we proposed to used as a negative feedback.

This negative feedback was embedded into a collaborative task for which we measured its effect. We founded that providing feedback regulates the behavior of a human collaborative partner. First, the subject avoided breaking the instructions and second it gave a communicative signal that fomented an intervention. This means that subjects that first face a failure reacted with more positive behaviors towards the continuity of the task than the ones given neutral feedback.

---

[3] Statistical significance calculated by Pair t-test method for an experiment with a power of 0.9 for a sample of 15 per case.

As future work, we are currently analyzing the behavior of the subject in the whole task rather than first failure and we are working into improve the readability of the positive gestures in our robotic face: happy and surprise. We also plan to test different negative gestures feedback such as: angry, worry and fear. Even though there is evidence that negative gestures have a negative impact on the subject's mental state, for instance when the Negative Attitudes Towards the Robot (NARS) test is applied [22]. We hypothesize that negative gestures have a place in human-robot interaction that will have a positive effect on the task and the mental state of the subject.

Acknowledgment. This research was financed by PCIC-UNAM, PASPA program and by CONACyT project 178673, PAPIIT-UNAM project IN107513 and CIDI, UNAM.

References

1. Breazeal, C.: Functions meets style: insights from emotion theory applied to HRI. IEEE Trans. Man Cybern. Syst. Part C **34**(2), 187–194 (2003)
2. Marsella, S., Gratch, J., Petta, P.: Computational models of emotion. In: A Blueprint for Affective Computing - A Sourcebook and Manual, vol. 11(1), pp. 21–46 (2010)
3. Knapp, M., Hall, J., Horgan, T.: Nonverbal Communication in Human Interaction, 8th edn. (2012). Cengage Learning
4. Fong, T., Nourbakhsh, I., Dautenhahn, K.: A Survey of Socially Interactive Robots: Concepts, Design, and Applications. The Robotics Institute, Pittsburgh (2002)
5. Arkin, R., Moshkina, L.: Affect in human-robot interaction. In: Calvo, R.D., ed. The Oxford Handbook of Affective Computing (2014)
6. Scheutz, M., Schermerhorn, P., Kramer, J.: The utility of affect expression in natural language interactions in joint human-robot tasks. In: HRI 2006 Proceedings of the 1st ACM SIGCHI/SIGART, Salt Lake City, Utah, USA, pp. 226–233. ACM (2006)
7. Breazeal, C., Kidd, C., Lockerd Thomaz, A., Hoffman, G., Berlin, M.: Effects of nonverbal communication on efficiency and robustness in human-robot teamwork. In: 2005 IEEE/RSJ International Conference on Intelligent Robots and Systems, (IROS 2005), Alberta, Canada, pp. 708–713. IEEE (2005)
8. Samer, A., Gabriel, S.: Perception of gaze direction for situated interaction. In: 4th Workshop on Eye Gaze in Intelligent Human Machine Interaction. DiVa, Santa Monica (2012)
9. Boucher, J., Pattacini, U., Lelong, A., Bailly, G., Elisei, F., Fagel, S.: I reach faster when I see you look: gaze effects in human–human and human–robot face-to-face cooperation. Front. Neurorobotic **6**(3), 1–11 (2012)
10. Nieuwenhuisen, M., Behnke, S.: Human-like interaction skills for the mobile communication robot robotinho. Int. J. Soc. Robot. **5**, 549–561 (2013)
11. Kondo, Y., Takemura, K., Takamatsu, J., Ogasawara, T.: A gesture-centric android system for multy-party human-robot interaction. J. Hum. Robot Interact. **2**(1), 133–151 (2013)
12. Haddadi, A., Crof, E., Gleeson, B., MacLean, K., Alcazar, J.: Analysis of task-based gestures in human-robot interaction. In: IEEE International Conference on Robotics and Automation (ICRA) Karlsruhe, Germany, pp. 2138–2144. IEEE (2013)

13. Seib, V., Giesen, J., Grüntjens, D., Paulus, D.: Enhancing human-robot interaction by a robot face with facial expressions and synchronized lip movements. In: En V. S.-U. Agency (ed.) Communication Papers Proceedings: 21st International Conference in Central Europe on Computer Graphics, Visualization and Computer Vision in co-operation with EUROGRAPHICS Association, pp. 70–77 (2013)
14. Eastwood, J., Smilek, D., Merikle, P.: Negative facial expression captures attention and disrupts performance. Percept. Psychophys. **3**, 352–358 (2003)
15. Midden, C., Ham, J.: Using negative and positive social feedback from a robotic agent to save energy. In: Persuasive 2009 Proceedings of the 4th International Conference on Persuasive Technology. ACM (2009)
16. Ekman, P., Friesen, W.: Unmasking the Face: A Guide to Recognizing Emotions from Facial. Malor Books, Los Altos (2003)
17. Pineda, L.A., Rodríguez, A., Fuentes, G., Rascon, C., Meza, I.: Concept and functional structure of a service robot. Int. J. Adv. Robot. Syst. **12**, 6 (2015)
18. Bennett, C.C., Šabanović, S.: Deriving minimal features for human-like facial expressions in robotic faces. Int. J. Soc. Robot. **6**(3), 367–381 (2014)
19. Berns, K., Hirth, J.: Control of facial expressions of the humanoid robot head ROMAN. In: RSJ International Conference on Intelligent Robots and Systems (IROS), pp. 3119–3124. IEEE (2006)
20. Miwa, H., Itoh, K., Matsumoto, M., Zecca, M., Takanobu, H., Rocella, S., Carrozza, M.C., Dario, P., Takanishi, A.: Various emotional expressions with emotion expression humanoid robot WE-4RII. In: Technical Exhibition Based Conference on Robotics and Automation, pp. 35–36. IEEE (2004)
21. Eastwood, J., Smilek, D., Mer, P.: Differential attentional guidance by unattended faces expressing positive and negative emotion. Percept. Psychophys. **6**, 1004–1013 (2001)
22. Nomura, T., Kanda, T., Suzuki, T.: Experimental investigation into influence of negative attitudes toward robots on human–robot interaction. AI Soc. **20**(2), 138–150 (2006)

Introducing a Methodological Approach to Evaluate HRI from a Genuine Sociological Point of View

Diego Compagna[✉], Manuela Marquardt, and Ivo Boblan

Technische Universität Berlin, FG Regelungssysteme EN11, Einsteinufer 17,
10587 Berlin, Germany
{diego.compagna,ivo.boblan}@tu-berlin.de,
manuelamarquardt@gmx.de

Abstract. The evaluation of human-robot interaction (HRI) is still a major methodological challenge. Despite the interdisciplinary nature of the field, sociologically inspired contributions are still rare. This paper aims to introduce a theory-driven method according to a sociological interaction concept to evaluate HRI and identify aspects of successful and satisfying interaction experiences. It combines Harold Garfinkel's breaching experiments with a frame analysis inspired by Erving Goffman. Sociologically, the method relies on a definition of social interaction based on the symbolic interactionism paradigm.

Keywords: Breaching experiments · Ethnomethodology · Frame analysis · Symbolic interactionism · Creativity engineering

1 Introduction

The normative criteria for a successful interaction in the field of HRI usually relates to the interaction being pleasant for the human being involved – without specifying further how this pleasure is defined or measured. A common goal is to create an interaction experience similar to the interaction with other human beings. Reflecting these trends, two questions are of major interest: Firstly, which factors are important in evaluating an interaction experience (reflecting critically on the assumption that it should be fluid and smooth or as "natural" as possible)? And in direct connection with this, secondly, to what extent and in which situations is an interaction similar to human-human interaction preferred? With regard to this second question, the field of sociology – and especially sociological theory – is promising in identifying factors constitutive of human-human interactions and transferring them situationally to human-robot interactions.

In this paper, we want to present the outline of a theory-driven method for evaluating HRI, which can address both questions at once. To this end, we primarily consulted Erving Goffman's "Frame Analysis" [1, 8], presented in his "Microstudies on Social Interaction" [2, 3] combined with Har-old Garfinkel's "Ethnomethodology" and breaching experiments [4]. Both authors assume that every social interaction is shaped by situation-specific – and therefore context dependent – expectations. Goffman primarily focuses on how these expectations can kept stable over time, whereas Garfinkel is primarily interested in the mechanisms (or methods) that the interacting

© Springer International Publishing Switzerland 2016
J.T.K.V. Koh et al. (Eds.): Cultural Robotics 2015, LNAI 9549, pp. 55–64, 2016.
DOI: 10.1007/978-3-319-42945-8_5

entities use to negotiate an alignment of expectations and to elucidate to one another how their actions should be understood. Garfinkel used breaching experiments to determine these ethno-methods. Combining these with a frame analysis, we are endeavoring to develop a method that is independent of particular cultural contexts, as it operates within the culturally defined boundaries of what is commonly considered to be a functioning interaction. This is especially important in light of the fact that comparative studies for Europe and Japan have reached the conclusion that the concepts of robot agency, as well as an appropriate user-robot interaction, differ significantly [5–7].

To introduce our sociologically inspired methodological HRI evaluation approach, in Sect. 2 we present general assumptions regarding HRI from a sociological point of view. The field of biomimetic robotics is a research area that revolves around human-centered construction and design – which therefore provides a promising application context for our method. The general assumptions of biomimetic robotics are discussed in Sect. 3, whereupon we want to show the potential of sociologically inspired evaluation methods for HRI, comparing two studies that used similar methods and outlining the advantages (as well as probable disadvantages) of our proposed method, combining breaching experiments with frame analysis (Sect. 4). The paper closes with some concluding remarks (Sect. 5).

2   General Assumptions Regarding HRI from a Sociological Point of View

Most of the research on HRI adopts a psychological view and therefore selects –in sociological terms – a methodologically individualistic approach (see e.g. [10, 11], and most of the paper presented in [12]). These approaches focus on the individual and disregard socially constructed reality. Genuine sociological contributions – using models, definitions, or theories of social (inter-)action – are still rare (see e.g. [13, 14]).

We use an interactional approach that procedes from George Herbert Mead's concept of symbolically mediated interaction [15]. Interaction is symbolically mediated because the meaning of a symbol (e.g. what is the right gesture for a greeting) is negotiated in an interaction between ego and alter. The meaning of a symbol depends on the reaction of alter on an action of ego and is therefore determined ex post – although intersubjectively between the two subjects ego and alter. Social reality and social meaning are the effect of successful interactions between social actors, as a reaction of alter is always related to the prior action of ego and is what gives ego's action meaning. In the end the understanding I have of myself also derives from the way that I see myself from others' point of view.

Within the scope of sociological work and thought, adopted theories and approaches are connected with the genealogic motivation of criticizing Talcott Parsons' predominant Structural Functionalism from the late 1950s up to the early 1970s. The main goal was – and in part still is – to show how social order is (literally) created by individuals. A systemic view of society and social reality puts the individuals in the shadow of functional necessities, normative expectations, and predefined role sets. Garfinkel's and Goffman's theories opposed this view of society. Their strength relied in proponing theories that could explain the emergence of social order as a bottom-up phenomenon (and not the other way around). However, there are major differences between these two approaches. Garfinkel's assumes an extreme position with regard to the importance of the subject and his or her performance to establish a stabile social system of reference. Goffman on the other hand assumes a position in the middle ground between interactionist theories and system theories, insofar as he relies heavily upon the concept of role, role sets, role expectations and so on.

Garfinkel's approach, ethnomethodology, begins from the very radical assumption that in every interaction, social actors have to readjust their understandings, beliefs, etc. regarding a situation. What actors have in common are the methods that they use to achieve that goal. In contrast, Goffman builds his theory of the relevance of frames on the assumption that a certain amount of meaning is pre-established by prior interactions. Even if the meaning of actions, words, and other symbols relies on the actors' re-affirmation, the main effort for them is to learn the specific codes of the culture they live in and interpret them correctly. The next goal is to determine the frame of reference in every given situation to quickly interpret the actions of alter ego successfully. The two theories are similar in a broader view of social theories. They both consider social reality as the result of the interpretation and affirmation (or establishment) of the meaning of actions with respect to words/symbols. They both rely on a constructivist view of sociality. In this regard, they are easily transferable to HRI settings. However, they represent one cornerstone of the possible assessment of a typical HRI situation. If HRI is seen as an equivalent to Human-Human-Interaction (HHI), these theories could be adopted to analyze the situation in terms of robots being participants in the construction of social reality. The benefit of using these theories would be to address the consequences implied by most social robot developments. To realize a HRI that is very similar to a HHI means inviting robots as equal social partners in the construction of what humans have (so far) claimed to be exclusively their product. On the other hand, using these theories could also reveal to what extent HRI is and should be similar to HHI. Adopting them to understand the details of HRI is the litmus test for the precise amount of genuine sociality involved. In newer concepts sociality isn't limited to humans. The agency of non-humans is taken into account in the Actor-Network-Theory, for example. Nonetheless, using two traditionally constructivist theories that focus purely on humans as social actors is still legitimate because (the development as well as the study of) social robotics is explicitly based on this very grounds.

These assumptions, presented here very briefly, are crucial for the question at stake: In terms of HRI, they become extremely important for knowing who or what could be an appropriate social actor and serve as alter for ego's interactions. Whether reality and identity can emerge from interaction depends on an entity's capability of being alter. This frame is of paramount importance in establishing a clear definition, especially when related to an interaction with a robot. As a human, alter is an entity that is usually seen as a fully social actor who has all the skills needed to recognize an action from ego as an offer. This capability provides a reaction that could be identified as the act of drawing a distinction. In this case, the interaction proceeds smoothly. If alter is a machine, ego will not expect to be able to create social meaning with it. If the robot's reaction is not the one I expected, it will not be able to alter my identity, beliefs, or definitions of reality.

As long as machines lack the ability to participate in symbolically mediated interaction, they won't be attributed the status of fully social actors.

In many cultures today, humans are the only entities who qualify as social actors [16, 17, 18]. This is not surprising in plural, complex modern societies, as it means a reduction of complexity to define entities capable of being social actors merely as humans. Otherwise ego would constantly have to decide if alter is a proper interaction partner and can react in an adequate way in order to build a common social reality, including identity and horizons of indisputable facts and meaningful questions. Gesa Lindemann states that social actors usually don't need a special indication or a referee to know which entity is suitable to interact with, because everyone knows that humans are valid social actors. For this reason, framing is vital in establishing an HRI setting that is suitable for analysis with breaching experiments: If the status is unambiguous and ego is fully aware that alter is just a robot or a machine, ego can deal with failures of communication and interaction flaws. If the status is not completely clear, if ego doesn't know whether the robot is just a machine (as would be clear for example with a washing machine) or if the robot actually possesses skills similar to human skills, the interaction is deeply disturbed. The possibility of implementing a breaching experiment is strictly linked to this observation and therefore to the correct framing of the whole situation. Conducting HRI research in different cultural contexts can benefit from breaching experiments with proper frame analyses, but only if symbols and triggers for crises are implemented cautiously [19].

3   General Assumptions Regarding a New Paradigm of Compliant Robots for Human-Centered HRI

The method presented here is highly suitable to be adopted with robots developed in the rather new paradigm of compliant robotics. This is because compliance is a central requirement for human-centered HRI and therefore approaches an HRI similar to HHI. Interesting approaches have come from the field of bionics or biomimetic robotics. The basic motivation behind the transfer of biological solutions to technological applications is the assumption that optimized biological structures have developed over the course of 3.8 billion years of evolution. To date, over 2.5 million different species have been identified and specific characteristics described to a great extent. Thus, in terms of biomimetics there is an enormous pool of ideas available for solutions to technical problems.

Biomimetics is the application of research and development approaches to technological applications that use knowledge gained from the analysis of living systems to find solutions to problems, create new inventions and innovations, and transfer this knowledge to technological systems. The idea of transferring biological principles to technology is the central element of biomimetics. A commonly accepted definition of biomimetics is:

"Biomimetics combine biology and technology with the goal of solving technical problems through the abstraction, transfer, and application of knowledge gained in interdisciplinary cooperation from biological models" [20].

A biomimetic solution usually includes several steps of abstraction and modification with regard to the specific biological solution at hand and is characterized by the creative transfer of ideas. The field is genuinely interdisciplinary, as biologists and engineers, as well as physicists and computer scientists are involved. In biomimetic robotics, biomimetic solutions are applied similarly in the design, control, and operation of robots. Potential applications are lightweight design, flow optimization, or animal-like behavior in navigation, motion control, and decision-making.

A common definition of a biomimetic robot is:

"A robot in which at least one dominant biological principle has been implemented and which is usually developed based on the biomimetic development process" [21].

The benefits gained from the use of biomimetic robots can be derived from inherent physical properties as well as from biomimetic-based "behavior." These two components complement one another. In biologically inspired robotics, some objects of study have proven to be particularly interesting. In these cases, the focus is on biological concepts (such as neural networks) or physical concepts (such as energy storage), depending on the initial perspective. These objects of study, further classified in terms of biological principles, include the following:

- Energy storage and recovery
- Structure and lightweight design
- Efficiency and power-to-weight ratio
- Neurobiomimetic feedback control
- Adaptive behavior /neural networks
- Sensor fusion
- Complex kinematic chains
- Protection /self-protection /protection of others

The mechanics of robotics deals with the motion of bodies (kinematics) and establishes relationships between the motions, mass, and forces acting on a body (dynamics). Current industrial robots are characterized by rigid kinematics that allow precise control using known methods. Since these robotic systems are powerful and fast, they fulfill their tasks very well in industrial environments, but they cannot be used at the present time for human-machine interaction without implementing additional measures. Properties such as strength, quickness, and precision lose their meaning in this case because the humans to which the robots need to adapt (and not vice-versa) are only moderately strong and move relatively slowly and with little precision. Instead, properties such as lightness and compliance (passive, active) become more important and are primary subjects of research at the present time.

The new paradigm is therefore promising in developing robots with whom HRI is not only successful, but satisfying. Following the assumption that HHI is the best reference for HRI one question to what extent the robot should be humanoid or humanized. One assumption is that the more biological principles are implemented in a biomimetic robot, the more the robot approaches its biological role model in its behavior and properties. As fully humanoid robots are not technologically possible at this point in time, an evaluation might be undertaken to evaluate certain human-like aspects. Examples might include a hand shake as a greeting or the passing of objects between a humanoid robot hand and a human hand. Possible influences in this scenario would be optical properties (e.g. five fingers) or haptic properties (e.g. compliance in the movements). One hypothesis that could be tested in HRI scenarios with breaching experiments and frame analysis is that the more similarities the robotic and the human hand have, the more successful and satisfying is the interaction.

4   Evaluating the Quality of HRI with Breaching Experiments

Several studies have made use of breaching experiments or similar experimental setups in HRI research [9, 22–29], but without developing a systematic approach with the goal of establishing a generic evaluation method for HRI. There are two main reasons behind breaching experiments' suitability for HRI settings. Firstly, breaching experiments operate on a high level in terms of the demands of social interaction. Secondly, the results aren't biased by social desirability – as is the problem with most survey and interview methods.

In ordinary HRI experiments, the experimentees are asked several questions after the performed interaction about the quality and subjective impressions of the interaction experience. Compared to HHI, HRI is often disappointing, as the experimentee has to carry out most of the interaction sequence. The human fills the gaps created by the robots' inabilities, which in contrast often leads to a more positive assessment of the experienced interaction. This is due to the fact that many experimentees try to emphasize their own efforts and frame their behavior as a response to the researchers' expectations.

With breaching experiments, one indicator for the success of an interaction is the adoption of repair strategies in case of flaws in interaction. These repairs can be observed and will only be adopted if the interaction is seen as worthy of repair.

These assumptions have to be combined with a situational frame: Muhl and Nagai [9] adopted a similar approach in their study and were able to identify six different strategies that experimentees adopted to repair an interaction with a robot. They used a deception strategy, which led to a reframing of the situation. In a laboratory experiment, they invited people to show objects to a robot and teach it some tasks related to these objects. The robot (or agent) was an animation on a screen and appeared as a baby face with eyes, eyelids, eyebrows, and mouth that could move expressively [30]. It had a biologically inspired saliency mechanism and followed the most important features in the scene with its gaze [31]. This is how the robot could display attention or address its human interaction partner without acoustic sensors or speech processing systems [9]. The deception strategy consisted in directing the robot's gaze temporarily in a wrong direction. Even though the interaction is rudimentary, the experimentees' repair strategies showed a certain belief in the robot's interaction capabilities. They tried to re-attract the robot's attention, for example by pointing to the object, showing it to the robot more closely, or making noise [9].

The interaction crisis was induced systematically. Ego applied cognitive framing about the state of its interaction partner alter, which is in turn a selection of how to approach alter. After a crisis arises, ego tries to repair the frame as long as it seems worthwhile. When realizing the breach, the frame is changed and no more attempts are made to repair the interaction.

Awareness of the frame is important in terms of another aspect: As a researcher, one can achieve a high degree of transparency by checking the realizability of the breaching experiment itself. Muhl and Nagai [9] showed that it is possible to achieve meaningful results in a laboratory setting. By comparing these results with research conducted in a stationary care facility for elderly people [14], Compagna & Muhl showed that the instrument can have different effects in different settings. In an everyday situation, people did not put forth any effort to repair an interaction sequence with a service robot. In contrast to this, a test person suffering from dementia went through the interaction sequence perfectly. A service robot had the task of serving a drink to the inhabitants, addressing the person by speaking [32, 33]. Most people didn't reply to the robot at all or preferred to talk to the other people present. Even if they accepted the drink and the robot thanked them, they didn't respond. This was due to the robot's inability to react flexibly, e.g. by turning a rejection into a request. In this situation, the reaction of alter cannot be semantically constructed as following an action of ego. Interaction in socio-logical terms is not only endangered by its possible failure, but doesn't occur at all. In contrast to these cases, the person with dementia saw the robot as a fully social actor and interaction partner, even if the robot failed to react or if the robot's reaction was not what it was expected to be.

Within the setting (and framing) of an everyday life situation, the adoption of breaching experiments constitutes a stricter test and should therefore only be imple-mented with care. As in the previously mentioned example, interaction experiments in a laboratory setting may be more fruitful, as the invited participants' motivation is higher and the situation is controlled – suppressing the tendency to address by-standers.

There should be no doubt that framing and context of an interaction are of paramount importance. Nevertheless, further research is needed to learn more about central aspects related to the frame in which the HRI is carried out.

5   Summary

The presented method of breaching experiments combined with frame analysis is highly suitable as an instrument in evaluating interactions between a human and a robot from a sociological point of view. A main indicator of the quality of the interaction is whether and how an explicitly induced crisis is repaired in a meaningful way. This approach can also highlight differences between individuals [9, 14] or broader cultural contexts.

Framing should also be reflected upon for the experimental setup. Findings of HRI breaching experiments have to be reanalyzed with respect to framing. If a proper frame analysis is conducted and the frame is set correctly, the findings can be very helpful in determining the quality of an interaction.

When meaningful repair strategies show that the robot is ascribed the status of a social actor, the interaction can be described "as if" it was a social interaction. One peculiarity of HRI is that humans tend to treat robots as if they were entities with the properties of social actors, although they may well be aware of the robots' inability to repair the interaction (one might also say that the robot is not able to process double contingency [16]). The analogy of interacting with a child could be helpful for further understanding the asymmetrical capabilities between the interacting entities. These differences in social interaction between humans and "as if"-social interaction between humans and robots need further consideration in future research. For the present state, we can assume that an "as if"-social interaction is of higher quality than an unsocial interaction.

In conclusion, observable repair strategies are a huge advantage of breaching experiments that can evidence the quality of the HRI, but only if the framing is considered and chosen wisely. Evaluation results can be used to further develop social robots with interactional skills.

Acknowledgments. The Research presented in this paper was primarily supported by the German Ministry of Education and Research. In addition, we would like to thank Mollie Hosmer-Dillard for her support and helpful comments.

References

1. Goffman, E.: Frame Analysis. Harper & Row, New York (1974)
2. Goffman, E.: Interaction Ritual: Essays on Face-to-Face Behavior. Anchor Books, New York (1967)
3. Goffman, E.: Relations in Public: Microstudies of the Public Order. Basic Books, New York (1971)
4. Garfinkel, H.: Studies in Ethnomethodology. Polity Press, Cambridge (1967)
5. Wagner, C.: Robotopia Nipponica: Recherchen zur Akzeptanz von Robotern in Japan. Tectum-Verl, Marburg (2013)
6. MacDorman, K.F., Vasudevan, S.K., Ho, C.-C.: Does Japan really have robot mania? Comparing attitudes by implicit and explicit measures. AI Soc. **23**(4), 485–510 (2009)
7. Kaplan, F.: Who is afraid of the humanoid? Investigating cultural differences in the acceptance of robots. Int. J. Humanoid Rob. **1**(03), 465–480 (2004)
8. Goffman, E.: Behavior in Public Places. Free Press, New York (1963)
9. Muhl, C., Nagai, Y.: Does disturbance discourage people from communicating with a robot? In: 16th IEEE International Symposium on Robot and Human Interactive Communication (RO-MAN), Jeju, Korea (2007)
10. Wykowska, A., Ryad, C., Mamun Al-Amin, M., Müller, H.J.: Implications of robot actions for human perception. How do we represent actions of the observed robots? Int. J. Soc. Rob. **6**(3), 357–366 (2014)
11. Feil-Seifer, D., Skinner, K., Matarić, M.J.: Benchmarks for evaluating socially assistive robotics. Interact. Stud. **8**(3), 423–439 (2007)
12. Herrmann, G. (ed.): Social Robotics. Lecture Notes in Computer Science, vol. 8239. Springer, Heidelberg (2013)
13. Burghart, C., Haeussling, R.: Evaluation criteria for human robot interaction. In: Proceedings of the Symposium on Robot Companions: Hard Problems and Open Challenges in Robot-Human Interaction, pp. 23–31 (2005)
14. Compagna, D., Muhl, C.: Mensch-Roboter-Interaktion – Status der technischen Entität, Kognitive (Des)Orientierung und Emergenzfunktion des Dritten. In: Stubbe, J., Töppel, M. (eds.) Muster und Verläufe der Mensch-Technik-Interaktivität, Band zum gleichnahmigen Workshop am 17./18. Juni 2011 in Berlin, Technical University Technology Studies, Working Papers, TUTS-WP-2-2012, Berlin, 19–34 (2012)

15. Mead, G.H.: Mind, Self, and Society. University of Chicago Press, Chicago (1934)
16. Luhmann, N.: Soziale Systeme: Grundriß einer allgemeinen Theorie. Frankfurt aM, Suhrkamp (1984)
17. Lindemann, G.: Doppelte Kontingenz und reflexive Anthropologie. Zeitschrift für Soziologie **28**(3), 165–181 (1999)
18. Baecker, D.: Who qualifies for communication? A systems perspective on human and other possibly intelligent beings taking part in the next society. Technikfolgenabschätzung - Theorie und Praxis **20**(1), 17–26 (2011)
19. Lutze, M., Brandenburg, S.: Do we need a new internet for elderly people? A cross-cultural investigation. In: Rau, P. (ed.) HCII 2013 and CCD 2013, Part II. LNCS, vol. 8024, pp. 441–450. Springer, Heidelberg (2013)
20. VDI: VDI Guideline 6220, Part 1: Biomimetics – Conception and Strategy. Differences between Biomimetic and Conventional Methods/Products. Verein Deutscher Ingenieure e.V., Düsseldorf (2012)
21. VDI: VDI Guideline 6222, Part 1: Biomimetics – Biomimetic Robots. Verein Deutscher Ingenieure e.V., Düsseldorf (2013)
22. Weiss, A., Bernhaupt, R., Tscheligi, M., Wollherr, D., Kuhnlenz, K., Buss, M.: A methodological variation for acceptance evaluation of human-robot interaction in public places. In: 2008 17th IEEE International Symposium on Robot and Human Interactive Communication, RO-MAN 2008, pp. 713–718. IEEE (2008)
23. Weiss, A., Igelsböck, J., Tscheligi, M., Bauer, A., Kühnlenz, K., Wollherr, D., Buss, M.: Robots asking for directions: the willingness of passers-by to support robots. In: Proceedings of the 5th ACM/IEEE International Conference on Human-Robot Interaction, pp. 23–30. IEEE Press (2010)
24. Sirkin, D., Mok, B., Yang, S., Ju, W.: Mechanical ottoman: How robotic furniture offers and withdraws support. In: Proceedings of the Tenth Annual ACM/IEEE International Conference on Human-Robot Interaction, pp. 11–18. ACM (2015)
25. Bauer, A., Klasing, K., Lidoris, G., Mühlbauer, Q., Rohrmüller, F., Sosnowski, S., Xu, T., Kühnlenz, K., Wollherr, D., Buss, M.: The autonomous city explorer: Towards natural human-robot interaction in urban environments. Int. J. Soc. Rob. **1**(2), 127–140 (2009)
26. Alac, M., Movellan, J., Tanaka, F.: When a robot is social: spatial arrangements and multimodal semiotic engagement in the practice of social robotics. Soc. Stud. Sci. **41**(6), 893–926 (2011). 0306312711420565
27. Nagai, Y., Rohlfing, K.J.: Can motionese tell infants and robots 'What to Imitate'? In: Proceedings of the 4th International Symposium on Imitation in Animals and Artifacts, pp. 299–306 (2007)
28. Short, E., Hart, J., Vu, v, Scassellati, B.: No Fair!!: an interaction with a cheating robot. In: 2010 5th ACM/IEEE International Conference on Human-Robot Interaction (HRI), pp. 219–26 (2010)
29. Takayama, L., Groom, V., Nass, C.: I'm sorry, dave: I'm afraid i won't do that: social aspects of human-agent conflict. In: Proceedings of the SIGCHI Conference on Human Factors in Computing Systems, pp. 2099–2108 (2009)
30. Ogino, M., Watanabe, A., Asada, M.: Mapping from facial expression to internal state based on intuitive parenting. In: Proceedings of the Sixth International Workshop on Epigenetic Robotics, pp. 182–183 (2006)
31. Nagai, Y., Asada, M., Hosoda, K.: Learning for joint attention helped by functional development. Adv. Rob. **20**(10), 1165–1181 (2006)

32. Compagna, D.: Reconfiguring the user: raising concerns over user-centered innovation. In: Proceedings VIII European Conference on Computing and Philosophy, pp. 332–336 (2010)
33. Compagna, D.: Lost in translation? The dilemma of alignment within participatory technology developments. Poiesis Prax. **9**(1–2), 125–143 (2012)

Robots as Participants in Culture

Head Orientation Behavior of Users and Durations in Playful Open-Ended Interactions with an Android Robot

Evgenios Vlachos[✉], Elizabeth Jochum, and Henrik Schärfe

Department of Communication and Psychology, Aalborg University,
Rendsburggade 14, 9000 Aalborg, Denmark
{evlachos,jochum,scharfe}@hum.aau.dk

Abstract. This paper presents the results of a field-experiment focused on the head orientation behavior of users in short-term dyadic interactions with an android (male) robot in a playful context, as well as on the duration of the interactions. The robotic trials took place in an art exhibition where participants approached the robot either in groups, or alone, and were let free to either engage, or not in conversation. Our initial hypothesis that participants in groups would show increased rates of head turning behavior-since the turn-taking activity would include more participants-in contrast to those who came alone was not confirmed. Analysis of the results indicated that, on the one hand, gender did not play any significant role in head orientation, a behavior connected tightly to attention direction, and on the other hand, female participants have spent significantly more time with the robot than male participants. The findings suggest that androids have the ability to maintain the focus of attention during short-term interactions within a playful context, and that robots can be sufficiently studied in art settings. This study provides an insight on how users communicate with an android robot, and on how to design meaningful human robot social interaction for real life situations.

Keywords: Android robot · Gender · Duration · Head orientation · Social human-robot interaction

1 Introduction

During the last decade, we have witnessed the development of a range of android and social robots designed to improve wellness in the five following domains: healthcare, companionship, communication, entertainment, and education [1–5]. Each domain can be separated into groups of tasks, where each task requires specific properties from each robot. The ability to select the "right" robot for the "right" task is of central concern to robotics researchers, as the robotic properties should match the expected social norms and rules attached to the task.

Here, we present the results from an open-ended Human-Robot Interaction (HRI) experiment in a playful and 'social' real-world setting (the task), that of an art-exhibition, with an android (the robot). Even though, nonverbal behaviour in HRI is a topic of

© Springer International Publishing Switzerland 2016
J.T.K.V. Koh et al. (Eds.): Cultural Robotics 2015, LNAI 9549, pp. 67–77, 2016.
DOI: 10.1007/978-3-319-42945-8_6

research that has emerged only lately, it is gaining scientific ground with a very fast pace. We focus on the nonverbal signals of users during short-term interactions with an extremely human-like android robot, and particularly we are interested in investigating head orientation, one of the most reliable cues implying attention direction, and a deictic signal indicating the current focus of interest. Humans as well as the majority of primates tend to turn their head, and direct their attention towards objects, or regions of space that are of immediate interest to them [6]. Even though the eyes also provide a reliable cue to where another individual is currently directing his/her attention, this information could be extracted equally from other social cues such as body posture, pointing gestures, and/or head orientation [7]. The contribution of head orientation to overall gaze direction is on average 68.9 %, and the accuracy of focus of attention estimation based on head orientation data alone is 88.7 % [8]. Argyle and Cook [9] estimated that listeners in dyadic conversations look at the speaker for 75 % of the time, whereas the speaker only looks at the listener for 41 % of the time. In addition, in multi-party interactions, head orientation is more important than gaze to the conversational flow [10, 11]. The gaze behaviour of users towards robots, as well as the gaze behaviour of robots towards users in order to establish and maintain communication has been extensively studied in the past [12–14].

Our aim is to examine users' nonverbal leaked information through head orientation cues during short-term face to face interactions with an extremely human-like robot in an entertainment context. *We hypothesize that participants in groups will show increased rates of head turning behavior in contrast to those who came alone, since the turn-taking activity would include more participants.* Apart from that, *we also want to examine if gender affects head orientation during HRI.* The duration of HRI can also be a signal indicating users' focus of interest towards the android. Since the robot's gender is male, *we hypothesize that the duration of HRI for female participants would exceed the duration for male participants.* In the following sections we explain why cultural settings and entertainment contexts are relevant sites for conducting robotic research, describe our methodology, present our results along with their statistical analysis, and conclude with a discussion on our findings.

2   Robots as Cultural Participants

Unlike laboratory settings, or real-world settings, art exhibitions are liminal spaces that invite open-ended types of interaction. Similar to observational field studies, participants at art exhibitions are more in control of choosing whether, or not to interact with the robots on display, and choose their proximity and physical relation to the art works (in this case, robots) [15, 16]. In short, participants determine the level of interaction, and involvement. Finally, art exhibitions provide a loosely structured framework that invites a phenomenological gaze not typically associated with robotics, or HRI research, where robots are "expected" to perform in certain ways. In this case, the interaction is much more playful, which introduces an element of creativity, and imaginative play. Such an environment might be conducive for further research when evaluating morphology, aesthetic experience, perceptual capabilities, and user expectation in education and entertainment scenarios. After all, the term "robot" (robota in Czech) was first encountered in a theatrical stage (R.U.R - Rossum Universal Robots by the writer Karel Čapek) and was actually a dramatic character. Robots have traditionally dwelled in the field of art and theater as representations of human beings [17].

A number of art, or exhibition centers have been hosts to robotic experiments; an observational field study took place in ARS Electronica Center in Linz, Austria with the Geminoid HI-1 [18], experiments with the mobile robot KAPROS were conducted at the Tsukuba Art Museum [19], Robovie was evaluated during its stay at the Osaka Science Museum [20], and perception experiments with Geminoid-F took place during theatrical performances [21], and eleven Robox autonomous mobile platforms were installed at the Swiss National Exhibition Expo.02 for five months to validate various techniques and to gather long-term HRI data [22]. Additionally, the National Museum of Emerging Science and Innovation (Miraikan) features several humanoid robots on display acting as museum guides as well as interactive installations [23]. The permanent exhibition provides users with the opportunity to communicate and operate two android robots (Kodomoroid and Otonaroid) and gives researchers the chance to study the nature of the human-robot interactions. Interactive art installations, and exhibitions are increasingly a relevant site for conducting HRI research.

3   The Experiment

3.1   Stimuli

We used the naturalistic-looking android robot Geminoid-DK (Fig. 1) that interacted verbally through recorded speech played through off board speakers, and displayed corresponding facial expressions (synchronized mouth movements that simulate speech, head orientation, eye gaze, and emotive facial expressions). The Geminoids are teleoperated robots built after existing persons, and were developed as communication mediums to address issues surrounding telepresence, and self-representation [24, 25]. For this experiment, the robot was equipped with very subtle pre-programmed facial motions that invited an intimate encounter (mostly depicting head nods, eyebrow raises,

Fig. 1. The Geminoid-DK (left) set-up during the experiment.

neutral and mild positive emotions like happiness, and surprise) running on an autonomous mode, and with pre-scripted verbal commands and responses running on a Wizard-of-Oz mode (an operator was controlling them) [26]. The operator was kept hidden from the participants in a small space surrounded by black curtains behind the robot. Figure 2 depicts the trial area, the location of the robot, the location of the cameras, and the location of the operator.

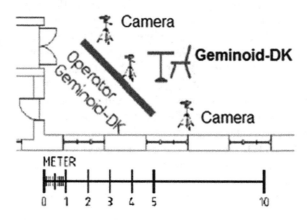

Fig. 2. The exhibition area, including robot, operator and camera location.

3.2 Design

The robot was exhibited as a part of a one-day art exhibition (4th of April, 2014) at Aalborg University, Denmark. The exhibition included another robot on display in a separate area of the exhibition space. The participants were not informed of the experiment in advance. Only when they reached the entrance of the exhibition was this information revealed to them. The participants were free to explore the exhibit for as long as they wished. Due to the amount of participants, their increased mobility in the exhibition area, and the open-ended nature of HRI we had three static video cameras recording at all times in order to always have a clear monitoring of the participant's head and body, as we were interested in recording clearly the head orientation behavior of those who choose to engage in HRI. One camera was mounted directly next to the robot (Fig. 1), one on the far- right side, and one on the far-left side, while all of them were adjusted to record at eye-level height of a seated person. Consent forms were on the table on the left side of the Geminoid-DK. Last, but not least, information concerning robot intelligence, abilities, and control mechanisms were deliberately concealed in order not to influence the participants.

As a change in the head orientation we consider a turn of the head to the right, to the left, upwards, or downwards as depicted in Fig. 3. Changes in head orientation due to change in body posture were also considered valid. Head tilting was not examined during this experiment, as well as head nodding and shaking since they are considered major modes of communication during listening turns signaling affirmation and negation [27].

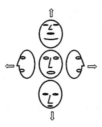

Fig. 3. Change in head orientation means rotation of the initial position of the head to the left, to the right, upwards or downwards. We examine 2 degrees of freedom (yaw, pitch).

3.3 Procedure

Participants were invited to interact with the robot simply by the mean of a single empty chair positioned directly in front of it, and could approach the robot either alone, or in small groups. In all the cases, the HRI was limited to one-on-one communication (one robot – one person). Participants attended the exhibition either accompanied by a group of friends and other persons, or alone. The HRI was categorized into five distinct phases: (i) Introduction, that included invitation and greeting phrases, (ii) Content, including phrases that would assist the robot to elicit information from the participants, (iii) Meta, with phrases that assisted in the progress of a dialogue, (iv) Consent form, where the robot asked participants to sign a consent form, and (v) Exit, with thankful and parting phrases. The scripted dialogue, along with the head movements and the facial expressions were more "natural" and realistic, rather than "artistic". The whole procedure and the scripted dialogue were organized to sustain communication for about 60 to 90 s. The language used was English. The number of head movements was labeled manually.

3.4 Participants

Most of the participants were university students and personnel from Aalborg University, non-native English speakers, but quite skillful with the particular language. We recorded 37 interactions, but we gathered only 33 signed consent forms, therefore we ended up with 33 valid interactions. From these 33 participants, 18 of them where male (54.5 %), and 15 (45.5 %) of them female. 21 participants came in groups (64 %), meaning that 21 participants were surrounded by other persons during the trial, but only one person from each group interacted with the robot while the rest of them were watching (12 males and 9 females). The remaining 12 participants came alone (36 %) meaning that during the interaction they were alone, and surrounded by none (6 males and 6 females).

4   Results and Statistical Analysis

4.1   Results

Male participants turned their head in a different direction during the HRI 7.4 times on average, while for female participants it was 9.4 times. The overall mean value was 8.3 head turns per participant. The most times a change in head direction was observed was 40 (female participant), while the least times was 0 (male participant). Participants who came alone turned their head 6 times on average (values ranged from 0 to 16). For participant who came in groups the mean value of change in head orientation was 9.6 times (values ranged from 2 to 40). Figure 4 depicts gender oriented mean values for both Group (8.5 for males, 11.1 for females) and Alone (5.1 for males, 7 for females) head turning behavior.

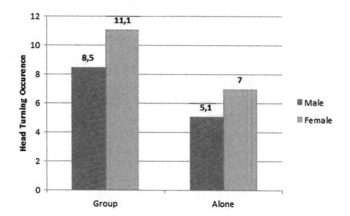

Fig. 4. Gender oriented mean values for head turning.

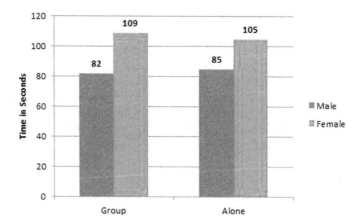

Fig. 5. Gender oriented mean duration of interactions.

The mean time of each interaction was 94 s, the longest one lasted 244 s (with a female participant), and the shortest one lasted 50 s (with a male participant). The mean duration of each interaction for participants within a group was 94 s, almost equal with the time spent by participants who came alone (95 s). The mean time for male participants was 83 s, and for female participants 108 s. Figure 5 illustrates gender specific mean times for both Group (male 82 s, females 109 s), and for Alone (males 85 s, females 105 s) interactions.

4.2 Statistical Analysis

An F-test (F = 2.954, num df = 20, denom df = 11, p-value = 0.068) indicated that the variances were equal, and a Two Sample 1-Tail T-test (t = 1.3319, df = 31, p-value = 0.096) showed that, since the p-value is marginal, *we cannot draw any strong conclusion related to head turning behavior and users approaching the robot in groups or alone* (could go either way). Figure 6 illustrates a boxplot for the Group - Alone variable.

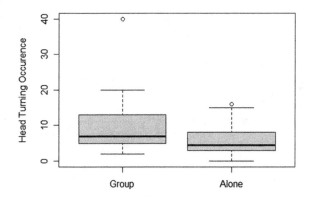

Fig. 6. Head Turning Occurrence Boxplot for Group (Median = 7, Standard Deviation = 8.499, Standard Error = 1.854) - Alone (Median = 4.5, Standard Deviation = 4.944, Standard Error = 1.427) variable.

In order to examine if gender played a significant role in head turning behavior, we performed an F-test (F = 2.95, num df = 14, denom df = 17, p-value = 0.036) which indicated that the variances were not homogeneous, and a Two Sample 2-Tail T-test (t = 0.7637, df = 31, p-value = 0.4508). According to the result, gender did not play any significant role in head turning behavior. Figure 7 depicts the boxplot for the Female–Male variable.

On the contrary, gender played a significant role in the mean duration of HRI. An F-test (F = 3.7998, num df = 13, denom df = 17, p-value = 0.01149) showed that the variances were equal, and a Two Sample 2-Tail T-test (t = 1.8902, df = 30, p-value = 0.03421) demonstrated that the mean duration of HRI for female participants was significantly longer from the mean duration of male participants. The Duration of Interaction Boxplot for Female-Male variable is depicted in Fig. 8.

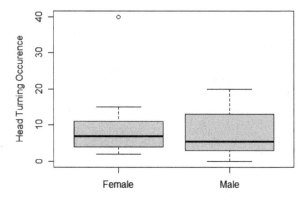

Fig. 7. Head Turning Occurrence Boxplot for Female (Median = 7, Standard Deviation = 9.485, Standard Error = 2.449) – Male (Median = 5.5, Standard Deviation = 5.522, Standard Error = 1.301) variable.

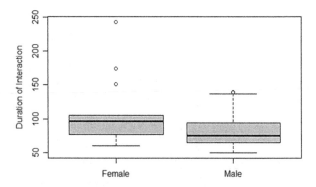

Fig. 8. Duration of Interaction Boxplot for Female (Median = 96.5, Standard Deviation = 49.619, Standard Error = 13.261) – Male (Median = 75, Standard Deviation = 25.454, Standard Error = 5.999) variable.

5   Discussion

Our initial hypothesis was that the head turning behavior of participants during dyadic interactions with an android in the "wild" during an art exhibition would be greater if they approached the robot in groups rather than alone. According to the statistical analysis, the relation between the head turning behavior of participants and the mode of approaching the robot (in groups, or alone) is not significant. Based on the data, we must reject our initial hypothesis that participants in groups would show increased rates of head turning behavior in contrast to those who are alone. Even in the presence of other visitors and bystanders, the turn-taking activity was limited to the human participant and the robot. In this study we also tried to investigate whether gender affected the head turning behavior, but our findings did not support this hypothesis. We did not find any correlation between gender, and the head orientation behavior of participants. However, the gender of the robot (male) might have caused an additional curiosity for the female participants to prolong their HRI. The statistical analysis, as well the mean times (83 s for males, and 108 s for females respectively) indicated that HRI for female participants lasted significantly longer than male participants. Our findings are consistent with previous research indicating that users are keener on rating opposite sex robots as more interesting, and convincing [28]. The mean duration of each interaction for both participants within a group, and for participants approaching alone was almost the same (94 and 95 s respectively), and it could mean that participants within a group were equally focused on the robotic interaction, and were not affected by their friends/surroundings. This could be an indication that participants attributed more human characteristics to the robot than mechanic. The fact that the scripted dialogue was more realistic and "natural", rather than artistic and abstract, might have also played a part in treating the robot more like human, rather than an object, or artifact.

Finally, we did discover that field studies such as art exhibitions, and gallery installations can be useful sites for studying HRI. These settings provide a platform sufficient for observing natural HRI in informal, "in the wild" settings. We plan to identify further aspects of human communication in order to isolate specific variables and study HRI in these types of settings. Future plans also include a follow-up study with the same robot in a less Wizard-of-Oz context to examine if participants will attribute less human characteristics to the robot, and another study with a female android robot, were we will replicate the experiment and examine if the gender of the robot indeed affects the mean duration of HRI for male, and female participants.

References

1. Broekens, J., Heerink, M., Rosendal, H.: Assistive social robots in elderly care: a review. Gerontechnology 8(2), 94–103 (2009)
2. Breazeal, C.: Social robots for health applications. In: Annual International Conference of the IEEE Engineering in Medicine and Biology Society, pp. 5368–5371 (2010)
3. Zhao, S.: Humanoid social robots as a medium of communication. New Media Soc. 8(3), 401–419 (2006). SAGE
4. Leite, I., Pereira, A., Martinho, C., Paiva, A.: Are emotional robots more fun to play with? In: The 17th IEEE International Symposium on Robot and Human Interactive Communication, RO-MAN 2008, pp. 77–82. IEEE Press (2008)
5. Abildgaard, J.R., Scharfe, H.: A geminoid as lecturer. In: Ge, S.S., Khatib, O., Cabibihan, J.J., Simmons, R., Williams, M.A. (eds.) ICSR 2012. CCIS, vol. 7621, pp. 408–417. Springer, Berlin, Heidelberg (2012)
6. Langton, S.R.: The mutual influence of gaze and head orientation in the analysis of social attention direction. Q. J. Exp. Psychol. Sect. A 53(3), 825–845 (2000). Routledge
7. Langton, S.R., Watt, R.J., Bruce, V.: Do the eyes have it? cues to the direction of social attention. Trends Cogn. Sci. 4(2), 50–59 (2000)
8. Stiefelhagen, R., Zhu, J.: Head orientation and gaze direction in meetings. In: CHI 2002, Extended Abstracts on Human Factors in Computing Systems, pp. 858–859. ACM (2002)
9. Argyle, M., Cook, M.: Gaze and Mutual Gaze. Cambridge University Press, Cambridge (1976)

10. Loomis, J.M., Kelly, J.W., Pusch, M., Bailenson, J.N., Beall, A.C.: Psychophysics of perceiving eye and head direction with peripheral vision: implications for the dynamics of eye gaze behaviour. Percept. **37**, 1443–1457 (2008)
11. Battersby, S.A., Healey, P.G.: Head and hand movements in the orchestration of dialogue. In: Thirty-Second Annual Conference of the Cognitive Science Society, pp. 1998–2003 (2010)
12. Imai, M., Kanda, T., Ono, T., Ishiguro, H., Mase, K.: Robot mediated round table: Analysis of the effect of robot's gaze. In: The 11th IEEE International Workshop on Robot and Human Interactive Communication, RO-MAN 2011, pp. 411–416. IEEE Press (2002)
13. Mutlu, B., Shiwa, T., Kanda, T., Ishiguro, H., Hagita, N.: Footing in human-robot conversations: how robots might shape participant roles using gaze cues. In: Proceedings of the 4th ACM/IEEE International. Conference on Human Robot Interaction, pp. 61–68. ACM (2009)
14. Minato, T., Shimada, M., Ishiguro, H., Itakura, S.: Development of an android robot for studying human-robot interaction. In: Orchard, B., Yang, C., Moonis, A. (eds.) IEA/AIE 2004. LNCS, vol. 3029, pp. 424–434. Springer, Berlin, Heidelberg (2004)
15. Kroos, C., Herath, D.C.: Evoking agency: attention model and behavior control in a robotic art installation. Leonardo **45**(5), 401–407 (2012)
16. Velonaki, M., Scheding, S., Rye, D., Durrant-Whyte, H.: Shared spaces: media art, computing, and robotics. Comput. Entertain. (CIE) **6**(4), 51 (2008). ACM
17. Demers, L.P., Horakova, J.: Anthropocentrism and the Staging of Robots. In: Adams, R., Gibson, S., Arisona, S.M. (eds.) Transdisciplinary Digital Art. Sound, Vision and the New Screen. CCIS, vol. 7, pp. 434–450. Springer, Berlin, Heidelberg (2008)
18. von der Putten, A.M., Kramer, N.C., Becker-Asano, C., Ishiguro, H.: An android in the field. In: Proceedings of the 6th ACM/IEEE International Conference on Human-Robot Interaction, HRI 2011, pp. 283–284. IEEE press (2011)
19. Maeyama, S., Yuta, S. I., Harada, A.: Mobile robots in art museum for remote appreciation via internet. In: Proceedings of the IEEE/RSJ IROS 2002 Workshop on Robots in Exhibitions, IROS 2002. IEEE Press (2002)
20. Shiomi, M., Kanda, T., Ishiguro, H., Hagita, N.: Interactive humanoid robots for a science museum. In: Proceedings of the 1st ACM SIGCHI/SIGART Conference on Human-Robot Interaction, HRI 2006, pp. 305–312. ACM, New York (2006)
21. Ogawa, K., Taura, K., Ishiguro, H.: Possibilities of Androids as poetry-reciting agent. In: The 21st IEEE International Symposium on Robot and Human Interactive Communication, RO-MAN 2012, pp. 565–570. IEEE press (2012)
22. Siegwart, R., Arras, K.O., Bouabdallah, S., Burnier, D., Froidevaux, G., Greppin, X., Tomatis, N.: Robox at expo. 02: a large-scale installation of personal robots. Robot. Auton. Syst. **42**(3), 203–222 (2003)
23. National Museum of Emerging Science and Innovation (Miraikan). http://www.miraikan.jst.go.jp/en/exhibition/future/robot/android.html. Accessed 15 June 2015
24. Vlachos, E., Schärfe, H.: The geminoid reality. In: Stephanidis, C. (ed.) HCII 2013, Part II. CCIS, vol. 374, pp. 621–625. Springer, Heidelberg (2013)
25. Vlachos, E., Schärfe, H.: Towards designing android faces after actual humans. In: Jezic, G., Howlett, R.J., Jain, L.C. (eds.) KES-AMSTA 2015. Smart Innovation, Systems and Technologies, vol. 38, pp. 109–119. Springer International Publishing, Heidelberg (2015)
26. Vlachos, E., Schärfe, H.: Android emotions revealed. In: Ge, S.S., Khatib, O., Cabibihan, J.J., Simmons, R., Williams, M.A. (eds.) ICSR 2012. LNCS (LNAI), vol. 7621, pp. 56–65. Springer, Berlin Heidelberg (2012)

27. Hadar, U., Steiner, T.J., Rose, F.C.: Head movement during listening turns in conversation. J. Nonverbal Behav. **9**(4), 214–228 (1985)
28. Siegel, M., Breazeal, C., Norton, M.: Persuasive robotics: the influence of robot gender on human behavior. In: The IEEE/RSJ International Conference on Intelligent Robots and Systems, IROS 2009, pp. 2563–2568. IEEE Press (2009)

Towards the Design of Robots Inspired in Ancient Cultures as Educational Tools

Christian Penaloza$^{(\boxtimes)}$, Cesar Lucho, and Francisco Cuellar

Pontificia Universidad Catolica del Peru (PUCP), Lima, Peru
{cpenaloza,cesar.lucho,cuellar.ff}@pucp.pe

Abstract. The use of robots as educational tools has demonstrated to be highly effective for attracting students to science and technology related academic fields. Although these academic fields are very important, we believe that other subjects such as language, music, arts, literature, history, etc., are also essential for future generations. For this reason, the goal of this research is to explore the potential use of robots as educational tools for non-technology related fields such as history. We discuss an alternative approach for designing robots inspired in traits and characteristics of historical figures that play an important role in the topic to be studied. We provide several examples of conceptual designs of robots inspired in ancient gods or historical characters of Mesoamerican and South American cultures. We discuss how some of the traits of ancient gods and characters could serve as inspiration for the appearance design of commercial robots, and how these robots could be used in educational environments to attract the attention of students to learn about this history topic.

1 Introduction

Robots have been used around the world in workshops for children as educational tools for experimenting with concepts that go from mobile robots [1] to the acquisition of physics knowledge through programming robotic platforms [2]. These types of workshops have demonstrated to be highly effective in attracting the interest of children and increasing the achievement scores in an informal learning environment [3]. The most commonly preferred tool is the LEGO Robotic Kit created as the expression of constructivist learning [4]. One of the first institutions to use this robotic kit as an educational tool was the Carnegie Mellon University and the results of years of experience are highly satisfactory [5].

Until now, the main goal of the use of robots as educational tools has been focused on attracting students to science and technology related areas for majors such as physics, math, computer science, electrical and computer engineering [6]. Although these academic fields are important for the development and economic growth of a country, other academic fields such as language, music, arts, literature, history, etc., are also essential for creating an identity for future generations. We believe that science and technology should be used to generate knowledge about non-technical topics, in particular about culture since cultural

© Springer International Publishing Switzerland 2016
J.T.K.V. Koh et al. (Eds.): Cultural Robotics 2015, LNAI 9549, pp. 78–84, 2016.
DOI: 10.1007/978-3-319-42945-8_7

development is a source of creativity. For this reason, the goal of this research is to explore the potential use of robots as educational tools for non-technology related fields such as cultural studies and history. Our aim is to enhance and enrich the experience of learning, as well as to attract the interest of students so they can appreciate the cultural aspects of history through robotic platforms.

2   Alternative Approach for Robot Design

Robot appearance design plays an important role in how people perceive the robot and get attracted to them. So far, most robot designs are based on proposals of experts, such as Kismet developed by Breazel *et al.* [8], or human-like Geminoid developed by Prof. Ishiguro [7]. However, robot design can highly affect the perspective of the people who interact with, as suggested by Bartneck *et al.* [9] who investigated the negative attitudes toward robots that people have.

Instead of using an arbitrary design proposed by a particular research group or person, we suggest that robot design for educational purposes should take inspiration from the academic topic to be studied. For example, we present the concept of a robot design to be used as an educational tool for the academic field of ancient history. Although ancient history covers a wide variety of topics, we chose to design our first robotic prototypes to study the cultural aspects of ancient Mesoamerican and South American religious belief history. In this sense, we take some design concepts from famous characters that play an important role in this particular topic, and use these concepts as inspiration for the design of the robot appearance, as detailed in the following section.

3   Ancient Cultural Based Robot Design

Mesoamerican religious belief history is perhaps one of the crucial topics to understand the roots of cultural practices that influenced economic and military activities of ancient cultures such as the Maya or Aztecs in the period that lies within the late pre-classic to early classic period (400 BC 600 AD) of Mesoamerican chronology [10]. Some of those cultural practices and even territorial domains still are alive today in territories such as the southern part of Mexico and Central America.

In ancient Mesoamerican beliefs, the existence of an extensive and complex array of deities and gods not only explained the origins of the world but also served as models for human behavior for commoners and elites alike [11]. We take the example of Quetzalcatl (pron. Quet-zal-co-atl), one of the most important gods in ancient Mesoamerica, whose name comes from the Nahuatl language meaning "feathered serpent" and was regarded as the god of winds and rain and as the creator of the world and mankind [11].

In order to take some of the concepts of Quetzalcoatl and portrait them into a robot design, we explored the actual meaning of the name which is a combination of the Nahuatl words for the *quetzal* - the emerald plumed bird Fig. 1b - and *coatl* - rattle snake Fig. 1c. Although there are some pictorial representations of

Fig. 1. (a) portrait from the Codex Telleriano-Remensis, (b) Emerald plumed bird, (c) Rattle snake

Quetzalcoatl that give more emphasis on the serpent characteristics, such as the one portraited in the Codex Telleriano-Remensis [12], shown in Fig. 1a, we also considered emphasizing the characteristics of the emerald plumed bird.

Figure 2 shows the conceptual design that portraits our representation of Quetzalcoatl as a flying robot (characteristic of the Quetzal bird) with its corresponding colors of green-yellow red that are also emphasized in the portrait of Codex Telleriano-Remensis. On the other hand, the head and tail of the robot are inspired on the rattle snake. This robot could be realized from a commercial flying robot such as Parrot AR.Drone [13], or in our case, using a custom-built drone robot. Nowadays there are commercial drone robots that are inexpensive and can be operated with a smartphone or tablet. Most importantly, they are easy to use even for people without technical background.

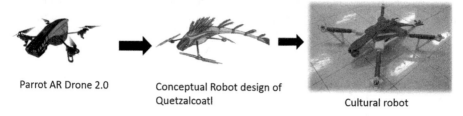

Parrot AR Drone 2.0   Conceptual Robot design of
Quetzalcoatl

Cultural robot

Fig. 2. Quetzalcoatl conceptual robot design (Color figure online)

In the same way, we can also take the example of Xbalanque (pron. X-balam-ke) one of the twins (along with Hunahpu) considered as a mythical ancestor to the Maya ruling lineages, and whose narrative is included in the famous book of Popol Vuh [14]. Figure 3a shows the commercial robot Nao from Aldebaran representing Xbalanque. The appearance takes into consideration the humanoid traits described in the Popol Vuh. On the other hand, the name of Xbalanque could be translated from the Maya language as 'Jaguar Deer' and thus could

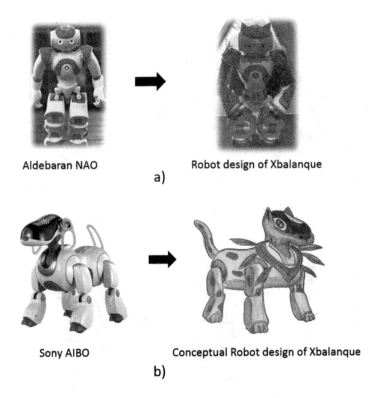

Aldebaran NAO   Robot design of Xbalanque

a)

Sony AIBO   Conceptual Robot design of Xbalanque

b)

Fig. 3. Xbalanque conceptual robot design

also be used as inspiration for the appearance of a robot as shown in Fig. 3b that uses as baseline the commercial robot AIBO from Sony company.

Apart from Mesoamerica, there are also numerous South American ancient cultures. One of the most representative from pre-Columbus chronology (1438–1533 AD) is the Inca culture. The number of gods of this Andean society is large, and many of them have an anthropomorphic appearance. In the Inca's beliefs, their gods were able to communicate with humans through their representations in stone, metal or wood which become alive figures [15]. As an example Wiracocha, identified as the most important god due to the fact that he brought "light into darkness", has seven eyes around his head that allowed him to watch everything around the world. He ordered the world and allocated the sun and the moon at the sky creating light, and then ordered humans to leave their caves. A representation of Wiracocha in stone is shown in Fig. 4, along with the conceptual design of a humanoid robot that poses the traits mentioned.

Although there are other historical characters that are representative of Inca culture, the king Pachacutec is perhaps one of the most well-known in the history. Pachacutec was the ninth Inca king of the Kingdom of Cusco which he transformed into the Inca Empire, and it is believed that the famous Inca site of Machu Picchu was built as an estate for Pachacutec [16]. The conceptual design

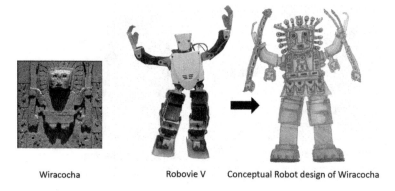

Wiracocha   Robovie V   Conceptual Robot design of Wiracocha

Fig. 4. Wiracocha conceptual robot design

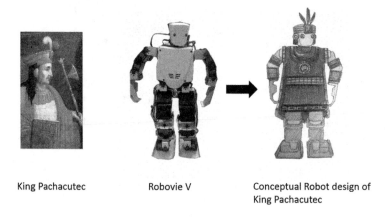

King Pachacutec   Robovie V   Conceptual Robot design of
King Pachacutec

Fig. 5. Panchacutec conceptual robot design

of a humanoid robot inspired in the traditional clothes used by Inca rulers of
that time is shown in Fig. 5.

4   Ancient Cultural Robot as Educational Tool

As shown in the previous section, all conceptual prototypes are based on commercial robots that could be easily acquired by academic institutions. These robots
can be operated with personalized application software installed in a smartphone
or tablet, and are easy to use even for people without technical background.
Moreover, the easy access to these robots would allow our methodology to be
widely used in schools or museums where the topics of ancient pre-Columbus
Mesoamerican and South American religious belief are first taught to students.

4.1 Learning by Building Robots

In this section, we propose a particular educational activity whose main goal is to allow students to learn about historic facts and characters in an interactive manner by applying the concept of robot design proposed in this paper.

Our approach consists of educational elements that have proved to be effective for educational purposes, such as team work, discussion, idea proposal and presentation. In particular the proposed activity is described as follows:

- Team Work - create several teams of students.
- Assign Character - assign each team a particular character of the ancient culture such as a god, king, etc.
- Learn and Discuss - allow each team to learn and discuss about the important characteristics or important facts about their assigned character.
- Propose - students should propose ideas of how the important traits could be used for the design of a robot.
- Do - Allow each team to adapt a commercial robot according to their proposed design.
- Present - Each team should make a presentation in front of the other teams about the ancient character assigned to them and the traits they chose to design the appearance of the robot.
- Demonstrate - Make a robot demonstration

Although the educational activity focuses on the learning of ancient religious beliefs of Mesoamerica and South America, this approach could be easily used for other academic subjects.

5 Conclusion

In this paper we proposed the potential use of robots as educational tools for non-technology related fields. Moreover, we proposed an alternative approach for designing robots inspired in traits and characteristics of historical figures that play an important role in the topic to be studied. As an example, we presented the conceptual design robots inspired in ancient gods and historical figures of Mesoamerican and South American cultures. We discussed how some of the traits of these ancient characters could serve as inspiration for the appearance design of commercial robots. Finally, we proposed an educational activity that will allow students to learn about the historic facts and characters in an interactive manner by applying the concept of robot design.

References

1. Jimenez, E., Caicedo, E., Bacca, E.: Tool for experimenting with concepts of mobile robotics as applied to childrens education. IEEE Trans. Educ. **53**(1), 88–95 (2010)
2. Williams, D., Ma, Y., Prejean, L., Lai, G., Ford, M.: Acquisition of physics content knowledge and scientific inquiry skills in a robotics summer camp. J. Res. Technol. Educ. **40**(2), 201–216 (2007)

3. Barker, B., Ansorge, J.: Robotics as means to increase achievement scores in an informal learning environment. J. Res. Technol. Educ. **39**(3), 229–243 (2007)
4. Papert, S.A.: Mindstorms: Children, Computers, and Powerful Ideas, 2nd edn. Basic Books, New York (1993)
5. Carnegie mellon robotics academy official webpage (2013). http://www.education.rec.ri.cmu.edu/
6. Ruiz-del-Solar, J.: Robotics-centered outreach activities: an integrated approach. IEEE Trans. Educ. **53**(1), 38–45 (2010)
7. Breazeal, C.: Designing Sociable Robots. MIT Press, Cambridge (2003)
8. Sakamoto, D., Ishiguro, H.: Geminoid: remote-controlled android system for studying human presence. Kansei Eng. Int. **8**(1), 3–9 (2009)
9. N. T. Bartneck C., S. T. Kanda T., y Kennusuke K.: Cultural differences in attitudes towards robots. In: Proceedings of the AISB Convention, HAtfield (2005)
10. Kent Reilly III, F.: Mesoamerican Religious Beliefs: The Practices and Practitioners. The Oxford Handbook of Mesoamerican Archaeology, November 2012
11. Taube, K.A.: Creation, Cosmology: Gods and Mythic Origins in Ancient Mesoamerica. The Oxford Handbook of Mesoamerican Archaeology, November 2012
12. Evans, S.T., Keber, E.Q.: Ritual, divination, and history in a pictorial aztec manuscript. Ethnohistory **44**(2), 419–420 (1997)
13. Parrot AR.Drone 2.0. http://www.ardrone2.parrot.com/
14. Coe, M.D.: Myth and image. In: Kerr, B., Kerr, J. (eds.) The Maya Vase Book: A Corpus of Rollout Photographs of Maya Vases, Justin Kerr (illus.), vol. 1, pp. 161–184. Kerr Associates, New York (1989)
15. Hampe Martinez, T.: Teodoro. Historia del Peru Lexus (2000). ISBN 9972-625-35-4
16. Rowe, J.: Machu Picchu a la luz de documentos de siglo XVI. Historia **16**(1), 139–154 (1990). Lima

Towards Socializing Non-anthropomorphic Robots by Harnessing Dancers' Kinesthetic Awareness

Petra Gemeinboeck[1(✉)] and Rob Saunders[2]

[1] Creative Robotics Lab, National Institute for Experimental Arts,
Faculty of Art and Design, University of New South Wales,
Sydney, NSW, Australia
petra@unsw.edu.au
[2] Design Lab, Faculty of Architecture, Design and Planning,
University of Sydney, Sydney, NSW, Australia
rob.saunders@sydney.edu.au

Abstract. This paper discusses a novel approach towards socializing non-anthropomorphic robots, which harnesses the expert knowledge of dancers to develop abstract robot morphologies and their capacity to move in affective and expressive ways. We argue that movement offers a key to socializing non-anthropomorphic robots. Our Performative Body Mapping (PBM) method investigates the possibility of using human movement experts to teach non-humanlike robots to move and interact. The paper outlines the conceptual framework of PBM and discusses an ongoing pilot study that engages professional dancers to study the relationship between abstract, simple morphologies and their potential to move in expressive, socially encoded ways.

Keywords: Social robotics · Creative robotics · Performance · Movement

1 Introduction

Robots are increasingly presented as 'social actors', designed to assist humans in therapy, eldercare, education and domestic tasks [8,24,26]. A 2013 study of the Japanese Ministry of Economy, Trade and Industry forecasted that by 2035, 50 % of total robot sales will be of service/personal robots that directly interact with humans [14]. Hence, the stakes for developing a better understanding of how to design socially competent machines are high.

Currently, the majority of research in Social Robotics and Human-Robot Interaction (HRI) focuses on anthropomorphic (humanoid) and zoomorphic robots [7,8,19]. The most well known example, emerging from MIT in the early 2000s is Breazeal's *Kismet*, a humanoid with controllable eyes, ears and lips that engages people in face-to-face interaction [5]. The underlying assumption is that robots that appear human– or pet–like are easier for people to relate

© Springer International Publishing Switzerland 2016
J.T.K.V. Koh et al. (Eds.): Cultural Robotics 2015, LNAI 9549, pp. 85–97, 2016.
DOI: 10.1007/978-3-319-42945-8_8

to [5,8]. Yet, humanoid or humanlike robots are technologically challenging and expensive to build [8,19], and studies consistently show that it is problematic if a robot's appearance and a person's expectation don't match. For example, the more humanlike a robot appears, the more people expect it to manifest human–level cognitive and social capabilities, leading to disappointing or frustrating interactions [8].

In this paper, we argue that movement can provide a key to socializing non-anthropomorphic robots. Studying the expressive qualities of movement and their potential to generate affect and empathy, rather than a robot's expressive physical features, opens up a much wider range of possible robot morphologies to design social agents. Furthermore, designs that don't imitate naturally existing agents allow for the robot's behavior to be the predominant factor for determining a person's attitude towards the machine without being biased by "preconceptions, expectations or anthropomorphic projections ... before any interactions have occurred" [8].

A key challenge when designing alternate robot morphologies and movements is to understand how an abstract or alien robot body can move and express itself in ways that humans can relate to. In the following we will introduce our research project that develops a novel approach to tackle this challenge by enlisting choreographers and dancers to harness both their movement expertise and embodied, kinesthetic understanding of how movement produces meaning and empathy.

The project is situated within the emergent cross-disciplinary area of Creative Robotics, which looks at human-robot interaction from a broad, culturally embedded perspective. The approach discussed here aims to open up uncharted territory with regards to a machine's kinesthetic abilities and how it can engender new aesthetic and affective experiences. The project is currently in its first development stages, and this paper will outline the conceptual framework and discuss the progress of a pilot study that engages professional dancers in a series of workshops to experiment with abstract machine morphologies and their potential for expressive movement.

2   Body Movement

Movement as a key element for developing a machine's expressive qualities has been explored by artists for more than 50 years. Important examples include pioneering works such as *The Senster* by Edward Ihnatowicz (1970) and Simon Penny's *Petit Mal* (1993). Discussing the latter, Penny talks about the "construction of a seemingly sentient and social machine ... an agent interface utilising purely kinesthetic or somatosensory modes which speak the language of the body and bypasses textual, verbal or iconic signs" [18]. Contemporary works exploring the affective potential of machine behaviours include Bill Vorn's *Hysterical Machines* (2006), Golan Levin's *Double-taker (Snout)* (2008), Mari Velonaki's *Fish–Bird* (2009) and the authors' *Accomplice* (2013).

Movement produces the kinesthetic sensations without which human agency, as characterized by action, cannot exist [17]. Far more than a matter of locomotion and physically interacting with the world, movement embodies culture and carries social meaning. According to Noland, it may require movement practitioners, expertly attuned to "the performing body's proprioceptive, kinesthetic, even affective experience of moving in prescribed ways", to understand to what extent movements and gestures "literally transform the bodies that perform them" [17].

As we will discuss in more detail below, at the core of our approach is the idea that, working with choreographers and dancers, we can develop a deeper understanding of how to cultivate kinesthetic relations between humans and non-familiar, abstract robot bodies. Our research puts forward an enactive approach to socializing robots and explores the concepts of corporeal literacy [3] and kinesthetic empathy [16,22]. The concept of corporeal literacy affords a perspective that recognizes the novelty of new embodied experiences while understanding that our bodies are cultured to both perform and perceive "in some ways rather than others" [3]. The interdisciplinary concept of kinesthetic empathy explores the affective potential of movement and, with it, our innate capacity to kinesthetically perceive other bodies. It is "a movement across and between bodies, which, in an artistic situation, can have affective impact with potential to change modes of perception and ways of knowing" [22]. This powerful connection has also been explored in interactions with objects and environments [16,22].

3   The Performative Body Mapping Method

Our project addresses two core open questions in HRI: (1) how should a sociable robot behave, and (2) how should it appear? Doing so, the research tackles two fundamental assumptions, namely, that a robot should interact with humans 'naturally' (i.e. in a recognizably 'human' manner), and that this is best facilitated if it appears humanlike [8,19]. Our hypothesis is that the expressive, dynamic and empathic qualities of movement can compensate for unfamiliar appearance in a robot's capacity to convey social agency. It is worthwhile noting here that, depending on the application, sociable robots may have very specific tasks that then define the main aspects of their appearance and behaviour. At this stage, our research responds to these questions and assumptions as a principle guiding our design and thinking about sociable robots and their affective potential. If movement is key to relating to and interpreting a robot, it could open up a much wider range of possible robot morphologies that are more cost-effective and adaptable to a changing social landscape than humanoid or pet-like morphologies.

At the center of our project is the development of the Performative Body Mapping (PBM) method for mediating between human and robot bodies. PBM places the robot's tactile–kinesthetic body and its movement at the center of meaning–making and eliciting affect to explore how non-humanlike robots can be taught to move and interact by human movement experts. The objective is for the robot to move according to its own abstract machine embodiment, whilst being 'seeded' with the movements qualities, textures and nuances that support social sense-making.

At the core of PBM is an autonomous robot with a non-humanlike, non-animal–like morphology with a capacity to learn how to move and a full–size, non-mechanical prototype of this robot body that becomes a 'costume' to be inhabited and/or animated by a dancer. The robot costume serves as an enabling constraint, an instrument for mapping between these two different bodies and their movement capacities, and for the robot to learn in a social, corporeal manner. It allows (1) for the dancer to learn embodying the machine body and to move with this unfamiliar body, and (2) for the robot to learn from the dancer by imitating the recorded movements from the dancer, disguised to mirror the robot's body.

3.1   Movement and Social Learning

In this project, the robot becomes the nonhuman apprentice of dancers who masquerade as the robot. Movement is at the center of social learning—learning from others. Dancers, for instance, 'sketch in dance' by "copying in real-time the movements of another dancer–the referent" [12]. The term 'sketching' also highlights that the copied movement will inevitably be a variation, due to differences in skill and body shape. In HRI, the most common type of social learning is imitation learning [1,9], used to teach robots humanlike skills and behaviors. Not surprisingly, a robot learning to copy a human requires mapping between entirely different embodiments, including different body shapes, sensorimotor capabilities, and movement repertoires [9], referred to as the *correspondence problem* [1]. Rather than focusing on learning a specific task, this project deploys imitation learning to capture the socially encoded, dynamic qualities of the dancers' movements. Using a costume that resembles the robot's body, a large amount of the morphological mapping between bodies is offloaded onto the dancer.

3.2   Computational Creativity

In addition to imitation learning, the robot will learn to explore and expand its movement abilities using a computational model of curiosity. The model is central to Computational Creativity, a sub-field of Artificial Intelligence that explicitly engages in questions of creativity. While its most common aim is to develop computational models of creative processes to study and support human creativity, researchers in computational creativity also produce autonomous systems capable of creative behaviors. Thus, for example, computational models of curiosity make it feasible for a robot to become an intrinsically motivated creative agent able to explore its own embodiment as well as its environment, where its reward is its learning as a result of this exploration [23]. This permits the development of artificial agents capable of proactively engaging with and learning to adapt to changing social scenarios [11].

4 Machine Movement Labs: A Pilot Study

In the following we discuss the progress of an ongoing pilot study, entitled Machine Movement Labs (MML), which engages professional dancers in a series of ten workshops to experiment with abstract morphologies and their movement capacity. The focus at this early stage is on challenging assumptions and preconceptions with regards to possible shapes and movements, rather than designing a robot with a specific social purpose in mind. More specifically, MML aims to explore how far we can push the relationship between abstract, simple morphologies and their potential to move in expressive socially-encoded ways. This open, exploratory approach allows us to explore a wide range of possible forms, materials, movements, and dramaturgical scenarios without the constraint of the robot design needing to fulfill a specific requirement.

4.1 Movement Strategy of BodyWeather

The pilot study engages three dancers from the De Quincey Company, including its artistic director, choreographer and dancer Tess de Quincey. De Quincey Co trains in BodyWeather, a practice founded on Butoh dance, which draws from both eastern and western dance, sports training, martial arts and theatre practice. BodyWeather uses images for the body to work from "to shift it out of its known, habitual pathways" [20]. The images, e.g., of external forces and their trajectories like wind or a pressure cooker, allow the dancer to escape the habitual and 'find' movements they wouldn't do otherwise. The body essentially moves in response to these imagined forces, sometimes multiple forces at once. De Quincey says "the whole point about BodyWeather is to go beyond the biomechanics through images [that is] we recruit the biomechanics to find new, unfamiliar ways to move" [20]. BodyWeather's kinesthetic empathy revolves around the body's sensitivity to and connectedness with its environment. Thus, while still bound to the human and socially encoded, BodyWeather dancers are already experts in finding other, non-habitual movements.

The objective of the workshops then is to explore the potential of dancers negotiating the expressive movement capacity of costume-like structures and objects, whereas it is now the costume that provides an external force for them to respond to and 'find' movements with. In the field of performance, the use of costumes to literally shape the performer's performance is not new. For his 1993 production of Tristan and Isolde, Heiner Mueller asked Yohji Yamamoto to design costumes for the singers "that would impede on the movement they are used to" [25].

4.2 Experiments (in Progress)

Our starting criteria for conceiving nonhuman morphologies were: no obvious front and back, no head or face, no limb-like structures. Another constraint for developing the costumes was that it can be reconstructed as a mechanical prototype capable of moving on its own. It is worth mentioning that the costumes

Fig. 1. Textile costume, inhabited by a dancer.

discussed below don't yet represent possible robot morphologies. Rather, at this early exploratory stage the objects used and developed serve as design probes to better understand the questions and issues involved, define the design criteria, and support the development of a language between the collaborators involved (artist, engineer, choreographer, dancer and costume designer).

In the first two labs we experimented with soft, textile structures, inhabited by the dancer, and surfaces with fiberglass rips to form architectural, parabolic shapes when bent, twisted and pulled by the dancers. However, the relatively soft shapes, requiring the dancer to give them a body (Fig. 1), turned out to be problematic: while the inhabitable forms could be richly animated with subtle movements, they were too reliant on the human body providing them with contour. The architecture-inspired, textile shapes, supported by elastic rips, produced interesting evolutions of geometric volumes but didn't allow for smaller, subtler expressions. It also seemed likely that the mechanical prototype would require large-scale mechanisms, external to the robot's body, to create the expressive shapes produced by the dancers.

Hence, for the third and fourth labs, we decided to work with simple costumes that formed a body on their own based on their material structure, but that could be transformed through the movements of a dancer inhabiting the structure. The first series of experiments also made clear that the simpler the shape, the more we could focus on the dancer's transformation of the body and its meaning, without being distracted by too many potentially moving parts. We experimented with a range of shapes and materials, and in the following will take a closer look at our experimentation with two of the most interesting 'objects'.

Fig. 2. Spiral tube costume (on the right) and textile tube with stiff plastic rings (on the left), both inhabited by a dancer.

Vertical Tube. The first object we experimented with was a spiral tube, 190 cm high with a 50 cm diameter, coated with a strong nylon fabric (Fig. 2). The tube acted as a relatively stiff spring that, by default, stood upright on its own, however could be compressed to a height of only 30 cm.

At first, the dancer physically engaged with the object and its materiality, exploring, testing, seeing and feeling what it can do and learning to negotiate its structural integrity. This included learning to move *with* the costume's structure by exploiting its ability to resist, transmit and transform forces applied to it. Soon the dancer (inside) began to improvise with the object, exploring different movement shapes, rhythms and their expressive qualities based on the feedback they received both from the object itself and the observers (the choreographer, another dancer, and the authors). The tube started swaying, barely noticeable and then with force, contracted in different parts, bent, crunched and twisted. The helical structure allowed for simultaneous contractions and expansions along the vertical axis of the object, as well as being bent as to produce multiple differently articulated planes pivoted along its core (Fig. 3). Both flexible and responsive, it enabled the dancers to effectively express themselves through tiny movements, a small swivel, teeter, twitch, or a crinkle here and there. Together with bigger gestures, either sustained or suddenly brought to a halt, this produced a very rich and affective performance.

We also built a 200 cm tube out of stiff plastic rings, strapped into an elastic scaffold and covered with a textile tube (Fig. 2), which produced a very different movement quality from the spiral tube. The springy spiral-shaped scaffolding proved more interesting, however, as it provided both a strong and flexible structural integrity. With an innate force to return to its default shape, it also allowed the dancer to apply force to transform the structure and, with it, its shape and expression. This play of tension proved to be very popular with the dancers.

Fig. 3. Spiral tube costume, showing multiple articulated planes pivoted along its core.

Box. We also experimented with perhaps the most obvious simple, abstract form, yet not the most apparent in terms of its evocative capacity—the box. The dancers were asked to inhabit and bring out the expressive potential of a $150 \times 55 \times 45$ cm cardboard box (Fig. 4). The stiff box shape got immediately interesting when it balanced precariously on edge or the dancer (inside) tipped it onto one corner. Tilting the box allowed for it to loose its stability and gravity and, with it, its 'boxiness', turning it into a strange, potentially fragile box-shaped character. To see the box move, sway and teeter as the dancers applied

Fig. 4. Box costume, tilted onto one edge.

different strengths of force and subtle variations of rhythm affirmed our belief that it is interesting and productive to have an expert dancer inhabit the strange body, rather than simulate the behaviors using a software-based model.

5   Discussion of Experiments

Not surprisingly, many of the affective qualities of movement, particularly with respect to their dynamic expression don't lend themselves to be captured in words, they exceed linguistic signification [15]. The affective power of movement, how it activates our body, happens before the cognitive process of language [21]. The empathic potential of this kinesthetic communication [10,21] is at the heart of our Performative Body Mapping approach as it aims to unlock the social potential of abstract non-anthropomorphic machines.

In one experiment, for example, the choreographer instructed the dancer inside the cardboard box to perform the abstract imagery of a question mark. When the dancer responded to the prompt, to us observers, the box took on a posture, overlaying notions of hesitation, inquiry and alertness. To be precise, however, rather than a posture, we had experienced the 'finding' of a movement, starting off with a hesitating twist that accelerated upwards, with a slight inclination, before it came to a sudden halt. This was not a visual representation of a question mark, but rather the bodily processing of what a question mark does, thus enabling us to feel the affective charge embedded in the box's gesture. Movement quality in dance concerns its dynamic, affective and expressive characteristics and always involves intentionality "articulated in and through" the movements. "Intentionality here does not refer to some kind of idea pre-existing the execution of the movement but rather describes the directionality and the distribution of intensity embodied within the movement and crucial to the quality" [4].

5.1   Animation vs. Performance

Parallel to performance, animation has a long and rich history of animating familiar but life-less shapes and objects and imbuing them with behaviors, disposition and intent. Similarly to our experiments, these objects can be surprisingly simple, as demonstrated in the classic example of Chuck Jones's *The Dot and the Line* (1965) or John Lasseter's *Luxo Jr.* (1986).

These animations are so successful because they commonly aim to anthropomorphize the object, imbuing it with a human character. Often, animators refer to the "personality of a character", conveyed through emotion, whereas the emotion is defined by the story. The 'readability' of the characters' actions relies on timing but also staging and anticipation. For example, "[i]n Luxo Jr., it was very important that the audience was looking in the right place at the right time" [13].

While animation techniques can be a very useful tool to develop a robot's movements, they have evolved in a very different medium, defined by its visual focus and the emotional impact of story telling. In contrast, robots are embodied objects, able to share and interact with our social environment in bodily ways. We can thus rely more on our kinesthetic sensitivities, without the need for the robot to be perceived as a humanized character. Our research aims to push this notion by investigating how we can utilize and train a machine's kinesthetic abilities for them to be readable by humans, without imbuing the machine with human personality. In particular, this pilot study investigates the affective kinesthetic abilities of different morphologies and materials.

Another important difference between animation and dance is in the aforementioned movement quality. Animation is about controlling the movement of a character, rather than 'finding' a movement or gesture and articulating intentionality in and through the dancer's body. Most computer animation systems use key frames to animate a character's movements. The animator defines poses, whose values are stored in key frames for the articulation controls of the character model, and the software interpolates between the values of these poses to render the full movement sequence [13].

Movement here doesn't emerge from the dancer embodying directionality and distribution of intensity but from externally defined, static poses, whose in-between is numerically interpolated rather than sustained, intensified or redirected. We can find an example in MIT's Interactive Theater, which deploys anemone-like robots capable of movements and behaviors that are readily apparent to the audience. As the theatre contains no dialogue, MIT's approach to animating the robot 'actors' was to transition between a set list of poses [6], rather than movements per se. Yet, much of what movement quality does, happens in-between and gets lost in an approach, which favors positionality over movement [2].

5.2   Concluding Reflections

This research into the potential of dancers training abstract, non-anthropomorphic robots is still at an early stage of development. In the first four workshops of our Machine Movement Labs pilot study, we have been able to experience three professional dancers moving, activating and transforming very simple objects, which, in turn, were able to trigger a range of affects and empathic responses. We are yet to develop autonomously moving mechanical prototypes and evaluate their kinesthetic performance in public settings to involve non-expert participants. Already at this early stage, as observers we found ourselves responding empathically to moving objects as abstract as a featureless tube or as stiff as a box (Fig. 5). They caused us to unwillingly lean our bodies with them, feel their subtle twitches, and to tense up when they threatened to fall. Based on these experiments, we found that kinesthetic empathy is not only a matter of us projecting onto the robot but also is a force that the moving robot body, despite it being radically different to our body, can actively transfer to us—make us feel.

The success of these first workshops attests to the potential of movement to turn an abstract object into an expressive, empathy–inducing social actor.

Fig. 5. Interaction with spiral tube (inhabited by dancer).

While we can't speak to the costume's potential in the robot's imitation learning process yet, we found that the costume plays a vital role in supporting the dancers mapping between the two bodies and developing an embodied understanding of what the robot body can do. In future workshops we will work with a costume designer to develop high-fidelity prototypes to explore the potential of dancers engaging with kinetic objects and transforming their intrinsic meanings in more detail. As part of the negotiation, we will develop a repertoire of meaningful movements, situated within the object's socio-cultural context. The findings of this exploratory study will support the next stage of developing the PBM method, beginning with the development of a 'mapping system' comprising physical and digital models and motion-capture interfaces to harness the dancer's knowledge and inform the first robotic prototype.

Interestingly, engaging with dancers in this Creative Robotics project not only provides us with insights into kinesthetic empathy and the material affect of movement. The dancers' approach and its deep entanglement with biomechanics, socio-cultural codes and empathy towards other, material agencies also expand our views of potential human-robot configurations. Research into the affective kinesthetic potential of abstract robot morphologies will not only lead to a novel approach for socializing abstract, non-anthropomorphic robots but will also provide a fertile ground for exploring new, culturally significant human-robot interactions.

Acknowledgment. The workshops discussed in this paper was supported by a University of New South Wales (UNSW) Research Grant (RG142730) and the ongoing research is supported under the Australian Research Council's Discovery Projects funding scheme (project number DP160104706).

References

1. Billard, A., Calinon, S., Dillmann, R., Schaal, S.: Robot programming by demonstration. In: Siciliano, B., Khatib, O. (eds.) Springer Handbook of Robotics, pp. 1371–94. Springer, Berlin (2008)
2. Bleeker, M.: Martin, massumi, and the matrix. In: Bleeker, M. (ed.) Anatomy Live: Performance and the Operating Theatre, pp. 151–164. Amsterdam University Press (2008)
3. Bleeker, M.: Corporeal literacy: new modes of embodied interaction in digital culture. In: Bay-Cheng, S., Kattenbelt, C., Lavender, A., Nelson, R. (eds.) Mapping Intermediality in Performance, pp. 38–43. Amsterdam University Press (2010)
4. Bleeker, M.: What if this were an archive? RTRSRCH J. **2**(2), 2–5 (2010)
5. Breazeal, C.: Designing Sociable Robots. MIT Press, Cambridge (2002)
6. Breazeal, C., Brooks, A., Gray, J., Hancher, M., McBean, J., Stiehl, D., Strickon, J.: Interactive theatre. Commun. ACM **46**(7), 76–85 (2003)
7. Castaneda, C., Suchman, L.: Robot visions. Soc. Stud. Sci. **44**(3), 315–341 (2013)
8. Dautenhahn, K.: Human-robot interaction. In: Soegaard, M., Dam, R.F. (eds.) Encyclopedia of Human-Computer Interaction, 2nd edn. Interaction Design Foundation, Aarhus (2013)
9. Dautenhahn, K., Nehaniv, C.L., Alissandrakis, A.: Learning by experience from others. In: Kühn, R., Menzel, R., Menzel, W., Ratsch, U., Richter, M., Stamatescu, I. (eds.) Adaptivity and Learning, pp. 217–421. Springer, Berlin (2003)
10. Foster, S.L.: Choreographing Empathy: Kinesthesia in Performance. Routledge, London (2010)
11. Gemeinboeck, P., Saunders, R.: Creative machine performance: computational creativity and robotic art. In: Maher, M., Veale, T., Saunders, R., Bown, O. (eds.) Proceedings of the 4th International Conference on Computational Creativity, pp. 215–219. University of Sydney, Australia (2013)
12. Kirsh, D.: Running it through the body. In: Proceedings of the 34th Annual Cognitive Science Society, pp. 593–598 (2012)
13. Lasseter, J.: Tricks to animating characters with a computer. Soc. Stud. Sci. **35**(2), 45–47 (2001)
14. Lechevalier, S., Nishimura, J., Storz, C.: Diversity in patterns of industry evolution. In: Proceedings of the 25th Annual EAEPE Conference (2013)
15. Manning, E.: Politics of Touch: Sense, Movement, Sovereignty. University of Minnesota Press, Minneapolis (2007)
16. McKinney, J.: Empathy and exchange: audience experiences of scenography. In: Reynolds, D., Reason, M. (eds.) Kinesthetic empathy in creative and cultural practices, pp. 219–236. Intellect Books (2012)
17. Noland, C.: Agency And Embodiment. Harvard University Press, Cambridge (2009)
18. Penny, S.: Embodied cultural agents: at the intersection of art, robotics and cognitive science. In: Socially Intelligent Agents: Papers From The AAAI Fall Symposium, pp. 103–105. Technical report FS-97-02. MIT, AAAI Press (1997). http://simonpenny.net/texts/embodied.html
19. Pfeifer, R., Bongard, J.: How the Body Shapes the Way We Think–A New View of Intelligence. MIT Press, Cambridge (2007)
20. de Quincey, T.: Video recording, Unpublished, 26 March 2015
21. Reynolds, D.: Kinesthetic empathy and the dance's body: from emotion to affect. In: Reynolds, D., Reason, M. (eds.) Kinesthetic empathy in creative and cultural practices, pp. 121–136. Intellect Books (2012)

22. Reynolds, D.: Kinesthetic engagement: embodied responses and intersubjectivity: introduction. In: Reynolds, D., Reason, M. (eds.) Kinesthetic empathy in creative and cultural practices, pp. 87–90. Intellect Books (2012)
23. Saunders, R.: Towards autonomous creative systems: a computational approach. Soc. Stud. Sci. **4**(3), 216–225 (2012)
24. Sharkey, N., Sharkey, A.: Living with robots. In: Wilks, Y. (ed.) Close Engagements With Artificial Companions. John Benjamins, Amsterdam (2010)
25. Suschke, S.: Müller macht Theater: Zehn Inszenierungen und ein Epilog. Theater der Zeit, Berlin (2003)
26. Weiss, A., Tscheligi, M.: Evaluating social acceptance and societal impact of robots. Soc. Stud. Sci. **2**(4), 345–346 (2010)

Robots and the Moving Camera in Cinema, Television and Digital Media

Chris Chesher[✉]

Digital Cultures Program, Department of Media and Communications, University of Sydney,
Sydney, NSW 2006, Australia
chris.chesher@sydney.edu.au

Abstract. The moving camera is a ubiquitous element in visual culture, and one that is undergoing significant change. Camera movement has traditionally been bound to the capabilities of human bodies and their physical equipment. Computer-based and robotic systems are enabling changes in image genres, extending the fields of perception for viewers. Motion control systems provide much tighter control over the movement of the camera in space and time. On television, wire-suspended cameras such as Skycam and Spidercam provide aerial perspectives above sports fields and music venues. Drones bring to the image a fusion of intimacy and magical elevation. An emerging domain of vision systems is in robotics and surveillance systems that remove the human operator entirely from the production and interpretation of images. In each of these cases, the question of the subjectivity and objectivity of images is complicated.

Keywords: Moving camera · Robotics · Motion control · UAVs

1   Introduction: Robot Controlled Cameras

Among the most widespread popular experiences with robotic technologies are the images made by robotically controlled cameras. These images are found in cinema, television, online media, and in robotics, and contribute to a range of new image genres. This paper provides a visual analysis of case studies of robot cameras that produce new kinds of moving image in a range of media. Theorists of the moving image make a distinction between a subjective camera, in which the perspective seems to belong to one of the characters in a scene, and an objective camera, which is more detached and abstracted [1]. Recent technologies are complicating the status of images as subjective or objective.

The robot camera, with its capacity for greater degrees of control than traditional camera operation, affords seamless transitions between subjective and objective viewpoints, allowing long-takes that move from intimate close-ups to wide-shots and back. Robotically controlled cameras using technologies such as industrial robots, motion control rigs and wire-suspended cameras and drones bring a variety of new resources to makers of moving images, allowing new forms of perception, and new kinds of subjective/objective positions. In movements, alongside edits, the robot camera tends to

© Springer International Publishing Switzerland 2016
J.T.K.V. Koh et al. (Eds.): Cultural Robotics 2015, LNAI 9549, pp. 98–106, 2016.
DOI: 10.1007/978-3-319-42945-8_9

enhance experiences of speed, vertigo and presence. These additions to cinematic language can be considered another domain of social robotics.

2   The Moving Camera in Cinema

The moving camera been a feature of cinema almost since its beginning in 1895. As early as 1897, the Lumière brothers attached a camera to a train to exploit the effect of movement. Camera movement became a powerful element in cinematic language. Jones [1] functionally groups the mechanisms by which a camera can be moved into five categories: tripod, dolly, jib/crane, handheld and Steadicam. The tripod serves to follow and reveal actors in a scene through tilting and panning. The dolly is effectively a tripod on wheels, which adds the dimension of movement forward and backward, and side to side. Jones argues that dolly moves present the camera as unaffected by the world: it moves in ways that are apparently outside the diegetic scene. The crane shot adds height and lateral movement, allowing movement in all three dimensions. This allows camera movements that are much more revealing: surveying a scene as closely or as widely as the director chooses.

Jones also discusses the hand-held camera, which allows free movement vertically and horizontally. However, because it is attached to the body of the camera operator, it tends to be shaky and unstable. This style of camera movement is often associated with a subjective view. Characters might directly address the camera as though the viewer is cast as a participant in the scene. Finally, the steadicam provides dampening mechanisms to deliver the high degree of fluidity of the dolly with the flexibility of the hand-held shot. Moving within the scene it is as transparent and beyond the physicality of the scene as the dolly. However, while it smooths movement, it remains attached to the body of the operator.

Recently there have been are some significant advances that add to Jones' categories of camera movement. The robotic camera is distinguished from other mediators of camera movement by its intensive use of computerised control, and its precise engineering. Motion control cameras extend Jones' typology by escaping the limits of human operation, offering the filmmaker the capacity to move the camera with greater reach, precision and repeatability. Where the operation of the dolly and crane are bound to capacities of the bodies of those on the set, robots are capable of faster, slower, and more precise movement. These are also designed to be integrated with virtual cameras in computer graphics.

The Cyclops camera mount from Mark Roberts Motion Control gives the filmmaker control of a full cinematic rhetoric of camera movements: track, lift, rotate, arm extension, head angle, pan, tilt, roll, camera, zoom, focus, iris as well as model movers, turntables and other peripheral devices [2]. With motion control systems, camera movement can be managed with computer-controlled precision. Many new visual effects become possible [2]. For example, by compositing multiple takes, elements can be made to appear or disappear from the scene while the camera moves smoothly. Another technique is crowd replication, where a small number of extras can be duplicated in different places across the shot and combined to create the appearance of a large crowd. Shooting against a green screen allows elements to be isolated and combined into a final shot.

As motion control systems can record the position and movement of the camera in 3D space with great accuracy, they can be combined seamlessly with computer graphics. The spatial features of the actual scene can be matched in the computer. Combining robotic cameras with calibrated 3D graphics supports the construction of complex composited sequences of movement. For example, in an early scene in *Moulin Rouge* [3] the camera appears to swoop across the roofs of Paris and descend into the window of the central character, Christian. The shot was composited from two takes: the virtual camera as it moves over the top of the 3D model of the Paris skyline, and the camera, as it moves towards the actor on a green screen stage. Because the movements were synchronised and scaled, the result is a seamless integration of the public space of the city and the private space of Christian's room. This gesture at once exposes the objective space of the city, and provides a subjective view of Christian as he imbibes an intoxicating sensory overload of music, movement, seduction and drugs.

Bordwell [4] critiques typologies such as Jones', observing that such accounts of camera movement concern profilmic processes: the movement of the camera in the production process. These approaches 'cannot specify the perceived screen event which we identify as camera movement' (20). The visual experience of movement for the viewer in the cinema is not reducible to how the camera moved during production, nor to formal logics that describe these moves. In animation in particular, camera movements can be complex constructions. 'Camera movement' does not necessarily involve cameras moving. Instead, Bordwell [4] argues that camera movement is best understood in terms of perception, and the camera movement effect. The spectator's experience is grounded in the spectator's physically passive position in the theatre. This creates the conditions for the viewer's perception of movement. Under these conditions, the spectator is able to experience the camera-movement effect, based upon the spatial cues in the images and sound. The viewer decodes the movement effect and reads off a sense of the spatiality of the world behind the screen. Camera movement helps spectators to establish an understanding of the spatial world of the scene (22). As the camera moves, the changing geometries in the scene reduce the ambiguities that can occur with the fixed camera: the 'kinetic depth effect' (23).

Thus we can hardly resist reading the camera-movement effect as a persuasive surrogate for our subjective movement through an objective world (23).

Deleuze [5] sees camera movement as critical to establishing what he refers to as the 'movement image' (24). Without a moving camera, the frontal point of view shows only an unchanging set without variation. It provides only defined 'immobile sections' (24) in space and time. The 'whole' of the moving image lacks change and lacks real duration. For Deleuze, the camera is central to the movement image. Deleuze's conception of the camera extends upon Bordwell's emphasis on perception to consciousness.

> ...we can say of the shot that it acts like a consciousness. But the sole cinematographic consciousness is not us, the spectator, nor the hero; it is the camera — sometimes human, sometimes inhuman or superhuman. ([5]: 20)

The superhuman camera has been a part of cinema for a long time. Even with traditional techniques, film-makers can deliberately disrupt the subjectivity of the camera, or the objectivity of the space. Bordwell gives the example of a 1927 film by Murnau called Sunrise: A Song of Two Humans [6]. The sets had been built using false perspective, so that foreground and background became distorted. A scene shot from the inside of a streetcar, with the still foreground contrasting with the moving background reflected the turmoil in the subjective relationship between the male protagonist and the vamp. Bordwell also points to scenes in other films in which the camera focusses on a character, tracks or pans away, and then shows the same character in different clothes. This use of off-screen space exploits the viewer's assumption that what he or she can see is homogeneous with what is visible, which of course it isn't.

Deleuze [5] argues that 'the mobile camera is like a general equivalent of all the means of locomotion that it shows or makes use of — aeroplane, car, boat, bicycle, foot, metro...' (22). Cinematic movement can take on the kinetic properties of any of these vehicles. The camera's movement can even extend beyond the capabilities of any known vehicle. For example, King Vidor's The Crowd [7] features a tracking shot that begins in a crowd, moves towards a skyscraper, climbing to the top of the building, selecting one of the windows, and then descending into a room full of rows of desks, until it focuses on just one (22). Deleuze also identifies two shots in Orson Welles' Citizen Kane [8] in which the camera moves from a building's exterior features and through a window into a large interior. These are distinctive because of the material constraints that they have overcome. They have a logic of 'as if': as if you climbed up the building, and down into its interior. But the smoothness of the ride makes it an impossible movement.

Related to these virtuoso shots is the practice of the long take: shots that last longer than what is considered normal in cinema. The long take typically features a sequence of events in a contiguous space that is revealed by a moving camera. Before the robot camera, the long take may be made by a long tracking or crane shot, steadicam or handheld. Directors such as Andrei Tarkovsky, Brian de Palma and Martin Scorsese are well known for this stylistic choice. Bruce Isaacs [9] analyses the work of director Alfonso Cuarón, who is well-known for his long takes. Isaacs identifies several philosophical and aesthetic functions of the long take: it provides a sense of realism that comes from the evidential legacy of the pro-filmic event. It asserts a material presence, such as is apparent when a liquid is splashed on the lens in the film Children of Men [10]. It challenges the dominant temporal and spatial regime of classical montage.

Recent technologies, including motion control systems, computer graphics, variable speed cameras, traditional special effects techniques, cinematic and televisual scenographic spaces can present disjunctive spatial and temporal changes in plain view. These techniques embody Deleuze's superhuman camera. They provide precise mediation of long take. An example is a promotional video for the digital film festival 'Resfest 2004' [11]. The commercial opens with the camera tilted upwards to reveal an inflatable wacky waving man, flopping around against a blue sky. The camera tracks to the right to show children holding puppets of military figures and professional wrestlers. The camera continues to track right, through a street scene, and into someone's kitchen. This person is attacked by vegetables. The absurdity continues across a number of locations, as the camera, smooth and indifferent, moves without faltering. What is notable in this piece during the relentless sequence of bizarre events is that the action takes place at a variety of depths: from close-ups to distant action, while retaining the same focal length.

Motion tracking is one way that the director can achieve a highly temporally saturated action in an apparently continuous *mise en scene* without any cuts.

Contemporary filmmaking often brings together multiple moving components to create the impression of a single scene. For example, the opening sequence in Martin Scorsese's Hugo [12] features a virtual camera floating over a detailed 3D model of Paris. The view drifts downwards and to the right. Gare Montparnasse comes into view, and the camera descends towards its entrance. The virtual camera moves along the digital platform, floating over an array of human figures, shot against a greenscreen, and composited into the space. The camera enters a cloud of steam, to mark the transition into a crowded film set. A large crane continues the inexorable forwards movement as extras move out of its way. Finally, the camera settles on the station's clock, with the face of the protagonist, Hugo, looking out of the number four. In spite of all its elements, the shot maintains a single movement image that propels the beginning of the film.

Another example of virtuosic use of robot cameras is the Cuarón film Gravity [13]. It convincingly presents a cinematic universe in micro-gravity. Early in the film a space mission is thrown into chaos when debris from an exploding satellite blasts through the space shuttle. At this point the central character, Ryan Stone (Sandra Bullock), is flung out of control on the Space Shuttle's robot arm, spinning uncontrollably. The arm swings her around towards the camera, revealing Stone's desperate face, and reflections in her visor. For this production, the company Bot & Dolly repurposed industrial robots with camera mounts and 3D software to allow the filmmaker to position the camera perspective within a 3D address space. With computers controlling both the rendered computer-generated space and the actual space it became possible for Cuarón and cinematographer Emmanuel Lubezki to create a sense of the space of orbiting around the earth.

To create this effect, the filmmakers mounted the camera on a robot arm, which moved around the stationary Sandra Bullock. This reversed movement helped create the impression that she was moving. Bullock was filmed inside a 'lightbox' that projected brightly lit images of her simulated surroundings onto her face and visor. The rest of the scene was rendered in computer graphics, with just the composited image of the face as actuality. The aesthetic of this robot-mediated scene gives both a sense of the sublime experience of space flight and a degree of intimacy with the astronauts under mortal threat.

The combination of CG, the robotic camera, and the lightbox create an image of the conditions of movement that are literally other-worldly. It conveys the experience of an inexorable inertia, which constantly threatens to send objects and bodies into the infinite depths of space. If a body starts spinning, it won't stop without being met with another force. This synchronised combination of digital and physical media combines multiple elements into a single composited space of movement.

3   Robot Cameras in Live Television

The cinema is not the only location for the reconfigured movement image. In live televised events, technologies such as the Spidercam and Skycam are wire-suspended cameras which provide fluid perspectives on sport, music and other events. The Spidercam consists of four motorised reels of cable that are strung from mounting points at high points at the corners of the playing field. The camera 'dolly' is suspended from these four cables. The reels are computer-controlled so that the camera can be moved at random to any point above the field. Spidercam has two operators. The pilot moves the camera around the field like using a computer game. The cameraperson controls the camera, zooming, focussing, tilting and panning.

The points of view available from Spidercam can range from intimately close to players to soaring wide shots. For example, in coverage of rugby league the camera will typically make a transition between a subjective view of players on the field to an over-view of the players from above, resembling the live diagram that a coach may have drawn to illustrate the movement of the players on the field. This capacity of the robot-mounted camera to move between the close-up of a subjective camera view to an objective aerial view in one shot distinguishes it from the traditional aerial shot. Conventions have developed for using Spidercam sparingly. It is typically used at kick-off, and in replays as the attacking team approaches the try line. It has also been used above the goal posts as the kicker kicks a conversion.

Skycam images in sport can increase viewers' sense of involvement in the action. A study by Cummins [14] showed that sport fans watching a sequence of plays in American college football using a subjective camera (the Skycam) felt a stronger sense of presence as engagement, and developed a better sense of spatial presence than those watching conventional television coverage.

Spidercam has also been used less judiciously at music events such as Eurovision. First seen in 2006 in Athens, the swooping Spidercam has become a mainstay of the coverage of this camp musical competition. Again, the Spidercam images are charac-terised by seamless transitions between subjective and objective viewpoints, moving from wide shots to close-up images of individual performers. At several points the camera circles around the stage, maintaining the framing on the performers, and giving viewers a stronger sense of the space of the performance. These images are no longer attached to the bodies of a camera crew, but instead move in between any-points-what-ever according to computer-controlled values.

4   Movement of Drones

The movement of a typical cinematographer's unmanned aerial vehicle (UAV), or drone, is smooth, continuous, computer-stabilised and energised. It is untethered, so can move to any location within the range of control and power. These features allow for the production of images with features that overlap with, but extend beyond, the norms of the television and cinema frames. It can shift rapidly between subjective and objective points of view. Its capacity for lift generates a sense of vertigo. It is what Stahl [15] refers to as 'drone vision' (659).

Drone vision is illustrated in the long take music video 'I Won't Let You Down' by OK Go! [16]. At one moment in the video, the camera is performing a standard tracking shot, moving backwards, away from the performers, when it suddenly seems to leap skyward. The figures on the ground recede to become abstracted points on the canvas of an empty car park. Opening and closing umbrellas, the accumulated figures become animated dots. Then the camera drops to retain its conventional relation to the performers' bodies before panning in a circle past all the people. Again, the camera launches itself vertically. Even more umbrella-wielding figures run into the open space and convert it into a pixelated screen, populated with text and images. Finally, the song ends, and the drone camera drifts above the clouds, away from the scene, revealing a cityscape and vanishing distance.

A more chilling drone camera sequence is the silent, monochrome YouTube clip 'Predator drone camera missile strike - missile cam', [17] which opens with the image of the roof of a non-descript building, apparently in the Middle East. At the centre of the image is a cross-hair. The image cuts to a wider view of the same building, at a different angle. Then another wide shot. The next shot is presumably the Predator taking off, searching for its target. Another wide shot. Then a missile cam view descends at the feet of two figures on the roof, and the image goes to noise. A wide shot again, showing the explosion. This narrative sequence, featuring a conversation of drones, and an act of violence traces the summary justice as enacted by human and non-human actors.

Franklin [18] observes that the visual field of war has changed over time. The still camera has been part of conflicts since the American Civil War. The World War Two newsreels privileged the war from the air. The lightweight cameras available during Vietnam allowed an intimate ground view of combat. Franklin argues that the first Gulf war saw a melding of the civilian gaze with the machine gaze in the weapon-cameras that suggested high technology, surgical and clean destruction. The steady perception of the drone camera has become commonplace in the early 21st century military image.

5   Robot-Readable World

Most of the examples of robot cameras in this paper so far are not strictly robotic. That is, they are remotely controlled or programmed by humans rather than being autonomous and reactive systems, even if they do rely on computers to operate. In this final Section I will look at the emergence of visual culture associated with cameras and other inputs that are part of autonomous systems.

A good current example is the autonomous car developed by Google. It uses a Velodyne laser range finder (LIDAR) to measure the distance of objects near the vehicle, up to a hundred metres away, and map the space around it. The car uses this data to avoid obstacles, and to match the onboard mapping information. The LIDAR data can also be converted into imagery: lines encircling the vehicle that indicate where laser reflections had taken place. Such imagery was used by the band Radiohead for the clip 'House of Cards' [19]. Autonomous vehicles can be considered as robotic camera mounts, or as curators of the moving image through the windscreen.

Another field of development in robotics is systems that interpret the content of live images and, use this information to control camera movement. This application is typically aimed for surveillance. For example, cognitive visual tracking [20] is engaged in 'the process of observing and understanding the behavior of a moving person' (457). It uses a variety of algorithms to identify the vectors of human movement. The system works by 'extracting semantic features of human motions and associating them with [a] concept hierarchy of actions' (458). Various systems then use one or more cameras to track one or more people within a defined space. A simple application will track people within the field of vision, while more sophisticated approaches will interpret whole 3D scenes or use conceptual analyses. For example, it could use face recognition to train cameras on an individual human moving through an entire area. The values in play for interpreting the scene are entirely instrumental, such as following someone to capture the best images for facial recognition.

The short video 'Robot-readable world' [21] edits together sequences of machine vision in action. Cars and people are identified with overlaid graphics. The variety of visual mark-ups through the clip indicates the diversity of applications at play in analysing everyday movement. Cameras count passing vehicles. Cars read road signs. Faces are recognised. Traffic is analysed. Movement in itself is lifted and abstracted from the raw video information, providing tactical data about changing vectoral spaces.

6   Conclusions

The increasing maturity of robotic systems promises to make available new resources for producers of visual culture. The tendency with these technologies is to complicate the subjectivity or objectivity of the image. They allow directors to move the camera from any location whatever to any other location. Autonomous camera platforms, from surveillance cameras to cinema and television, are likely to become 'smarter' and increasingly uncanny.

References

1. Jones, M.: Camera in motion part 2: means and methods. Screen Educ. **61**, 112–117 (2011). http://search.informit.com.au/documentSummary;dn=938208349941246;res=IELLCC
2. Mark Roberts Motion Control (n.d.) 'Cyclops | Mark Roberts Motion Control'. http://www.mrmoco.com/cranes-rigs/products/rigs/cyclops/. Accessed June 2015
3. Moulin Rouge, Director: Baz Luhrmann. Australia: Twentieth Century Fox and Bazmark Films (2001)
4. Bordwell, D.: Camera movement and cinematic space. Ciné- Tracts. J. Film Commun. Cult. Polit. **1**(2), 19–25 (1977)
5. Deleuze, G.: Cinema 1: The Movement Image. University of Minnesota Press, Minneapolis (1986)
6. Sunrise: a song of two humans, Director F.W. Murnau, Fox Film Corporation (1927)
7. The crowd. Director: King Vidor, USA: Metro-Goldwyn-Mayer (1928)
8. Citizen Kane. Director: Orson Welles. USA: RKO Radio Pictures and Mercury Productions (1941)
9. Isaacs, B.: "Reality Effects: The Ideology of the Long Take in the Cinema of Alfonso Cuaron". In: Leyda, J., Denson, S. (eds.) Post Cinema: Theorizing 21st Century Film. (Reframe Books) (forthcoming, 2015)
10. Children of Men. Director: Alfonso Cuarón. Universal Pictures, Strike Entertainment and Hit & Run Productions (2006)

11. RESFEST 2004 Festival Open Directors: Motion theory and kozyndan, Resfest. Vimeo: https://vimeo.com/356891 (2004)
12. Hugo. Director: Martin Scorsese USA: Paramount Pictures, G.K. Films and Infinitum Nihil (2011)
13. Gravity. Director: Alfonso Cuarón, USA Warner Brothers, Esperanto Filmoj and Heyday Films (2013)
14. Glenn, C.R.: The effects of subjective camera and fanship on viewers' experience of presence and perception of play in sports telecasts. J. Appl. Commun. Res. **37**(4), 374–396 (2009)
15. Stahl, R.: What the drone saw: the cultural optics of the unmanned war. Aust. J. Int. Aff. **67**(5), 659–674 (2013). doi:10.1080/10357718.2013.817526
16. I Won't Let you Down. OK Go. USA & Japan: Mori Inc
17. Super Terrific Happy Hour "Predator Drone Camera Missile Strike - Missile Cam". https://www.youtube.com/watch?v=KcEoRae6Yog (2014)
18. Bruce, F.H.: Vietnam and Other American Fantasies. University of Massachusetts Press, Amherst (2000)
19. House of cards. Radiohead. Directors: James Frost and Aaron Koblin, Radiohead (2008)
20. Bellotto, N., Benfold, B., Harland, H., Nagel, H.-H., Pirlo, N., Reid, I., Sommerlade, E., Zhao, C.: Computer vision and image understanding. Comput. Vis. Image Underst. **116**(3), 457–471 (2012). doi:10.1016/j.cviu.2011.09.011
21. Brown, M.: Robot readable world shows us life through the eyes of machines. Wired UK, 19 April 2012. http://www.wired.com/2012/04/robot-readable-world/

Robot-Supported Food Experiences
Exploring Aesthetic Plating with Design Prototypes

Christian Østergaard Laursen[2], Søren Pedersen[2], Timothy Merritt[1(✉)], and Ole Caprani[2]

[1] Department of Engineering, Aarhus University, Aarhus, Denmark
tmerrit@eng.au.dk

[2] Department of Computer Science, Aarhus University, Aarhus, Denmark

Abstract. Robots are increasingly taking up roles in society to support and interact with humans in various contexts including the home, health-care, production and assembly lines, among others. Much of the research focuses on efficiency, speed, accuracy of repetitive tasks, and in most cases the robot simply replaces and performs work tasks originally performed by humans. Looking beyond the simple replacement of humans with robotic servants, we focus on increasing creativity and pleasurable experiences supported by robots for the preparation, serving and consumption of food. This is a culturally rich area to design for, which is steeped in tradition, social norms and expectations. How can robots play a role in this context? We observed and interviewed chefs to gain a sense for opportunities for robotic technologies. We then created nine exploratory video prototypes involving food preparation with a robotic arm taking departure in themes of haute cuisine, "plating", and the arts in order to show some of the capabilities of robots and to spark their imagination for possible future uses of robots in the kitchen. Through questionnaires and interviews, we gained feedback from ten chefs with resulting themes including harsh criticism and resistance to robots as well as desire and interest for robots to support food experiences as a partner in the restaurant. We discuss emergent themes from the feedback and provide discussion on future work needed to explore robots as partners in creative contexts.

1 Introduction

Robots are finding roles as visible actors alongside people across society from supporting surgeons in the mission critical surgical theatre [35] to providing comfort as a virtual companion [15]. These roles go beyond the traditional view of robots acting as replacements to humans or acting as a servant, toiling away in the background. Even in the industries most heavily relying on robots for production, such as the automotive industry, there is an increasing shift in perspective toward a more collaborative approach with robots working with humans appearing in various forms [43]. Going beyond the focus on sequential, repetitive work, and simply replacing humans, there are examples emerging in the home and service sectors including food and beverage, in which robots take up supporting roles, acting

© Springer International Publishing Switzerland 2016
J.T.K.V. Koh et al. (Eds.): Cultural Robotics 2015, LNAI 9549, pp. 107–130, 2016.
DOI: 10.1007/978-3-319-42945-8_10

with some degree of social awareness [24]. Moreover, researchers have examined the close contact of humans and robots and the attitudes and feelings evoked by robot companions [45]. Researchers have signalled emerging and evolving culture and new attitudes around robots in society as not only passive artifacts, but as partners in human activities and creators of culture [38,40].

In the food industry, an ongoing concern has been that with more technology and standardization, there can be negative effects from deskilling chefs, staff reduction, reduced labor mobility, and job losses [36]. In a recent opinion report by the European Commission that reached out to more than 26,000 people across 27 member nations, there are wide differences in opinion within EU citizens regarding where robots are welcome in society and sectors in which people believe robots should be banned [8]. In that report, a common theme was identified in that robots were viewed as suitable for utilitarian purposes and dangerous environments, but not welcome in the more 'human' contexts such as caring for the elderly or taking care of children. While the survey is helpful identifying some of the popular opinions, it is unclear how people form their beliefs about the capabilities of robots. It is the aim of this paper to engage more closely with people in the food industry to gain a sense for their attitudes and opinions about robots and ways they imagine robots in the food industry. Our inquiry is inspired by the perspective of 'co-design' that aims to involve specialized users and develop solutions that fit into existing practices and uncover unmet needs [41]. We feel it is necessary to reach out to chefs to examine how we might co-design experiences for the future kitchen with robots, not to accelerate a reduction in the workforce, but to look for opportunities for new experiences and uses for robots as collaborators.

The focus of this paper is to explore and shed light on how robotic agents can be implemented in the modern gastronomical kitchen as a collaborator, assistant, or an extension of the chef. The paper is guided by the following important questions regarding human-robot interaction and collaboration: *How can robots support and enhance desirable experiences related to the preparation, serving, and consumption or sharing of food? How can the design of robot-supported experiences related to food benefit from existing knowledge, attitudes, and techniques of people in the food industry?*

The structure of the paper is as follows: we briefly highlight and discuss some of earlier and current work within the food industry and the research of robots supporting preparation of food, next we describe our research through design approach including the empirical work along with the development of the exploratory design prototypes. We then discuss the results of our work in terms of emergent concepts and ideas based on preliminary findings and impressions from chefs. We highlight two perspectives on the envisioned roles and visibility of robots and food experiences. Lastly, we present emerging ideas and highlight interesting future work.

2   Related Work

There is a long history of research and invention involving robots in the food industry. We review some of the ways robotic technologies have been introduced and relevant points raised by the research community. Of particular interest are the opportunities and challenges for human robot collaboration and research that has called for more culturally meaningful ways to involve technology and food experiences.

A cursory search through a patent database yields many examples of robots and automation technologies in the food industry. In the patent details for a robotic cooking system described in [14], patents describing automated tools for handling ingredients and preparing food were appearing as early as the 1920s. There are many examples of robots in the food industry which focus on efficiency and speed of repetitive tasks, packing vegetables [6], mixing components, and other tasks that would normally be performed by a human worker [46]. Contemporary manufacturers including Universal Robots, ABB and others advertise capabilities of their robots being able to deliver non-stop productivity with a reduced risk of employee injury by offloading repetitive tasks [1,7] - providing the benefit of productivity and safety to the production line.

There are various examples of technology designed around food including smart kitchens, augmented utensils, and design of culturally sensitive robot experiences with food [30]. There have been cooking and serving robots developed for assisted living facilities in which elderly people are confined to wheelchairs [31]. More recent examples include the use of robotic arms to assemble and cook ingredients for simple dishes [37] and more advanced robotic arms that can record the movements and techniques of the chef and make them available for replay through an online content store [33].

In terms of collaboration, there are examples in which researchers envision real-time cooperation with robots without relying on preplanned tasks. Shah et al. [42] presents a system for optimizing human-robot team performance by letting the robot more naturally emulate the decision processes of human teams. This is particularly important in the kitchen, which is a very dynamic context and the handling of food ingredients and dishes require great positional accuracy and context awareness [48]. There has been research on robots performing a variety of dynamic tasks within the kitchen, such as a task planning system for collaboration between robot and human [27] or recognizing human activity in a cooking context, in order for a robot to better support and guide future actions [25].

Researchers proposed metrics for human-robot interaction including performance measures, but also social and aesthetic experiences [44], yet there are few examples of research exploring robots and the creation of aesthetic experiences around food. Whether it concerns preparation or consumption, it is important to note the difference of solving existing problems with technology or augmenting current practises within the kitchen. There has been an increased interest in interactions related to food in the human computer interaction research community [16]. Grimes and Harper [28] suggested a shift in focus from corrective technologies to enhancing existing experiences and practices through more emotionally relevant measures. *CoDine* provides an inspiring example of robotic technologies designed to connect remote diners in order to share dining experiences and communicate in and around food [47]. We appreciate their focus on using robotics to support "...experience over efficiency and shared interaction over information."

3   Research Problem

While there are various examples of robots supporting the preparation and dispensing of food or drinks, the focus in much research has been largely on replacing the human chef or server to bring improvements of speed and efficiency. Our focus is not on solving current problems linked to efficiency, but instead our goal is to explore and imagine how robots can support aesthetic and pleasurable experiences with food, and support the creative process in the kitchen. Much of our daily life revolves around preparation of meals and the topic could be rigorously reviewed and explored in various directions. In this work, we are not interested in optimizing a known design or prototype, but rather to engage in a design exploration to find points of divergence and possibilities for future inquiry. To narrow the focus of the research we report here, we are guided by the question, *How can robots support chefs and serving staff in the design of experiences with food?*

4   Method

We adopted a research through design approach [26] in order to begin to explore the design space of robot-supported food experiences. We first immersed ourselves into the context of the kitchen with observations and interviews with chefs. We then developed simple experience prototypes involving the placement and preparation of food. We shared these with chefs and asked them to answer questionnaires and take part in co-design interview sessions to imagine possible uses for robots in supporting experiences with food.

Our investigation recognizes the difficulty in designing new technologies for situations in the real world. We do not want to over simplify the design space and we recognize that singular cases are not generalizable across cultures. In the context of food preparation and serving, there are many practices and concerns of chefs that are culturally significant and do not necessarily need to be changed. Our focus is not to remedy known problems, but to explore the design space and uncover potential opportunities for robotic technologies. As such, our inquiry is an attempt to empathize with the practices, struggles and concerns of chefs. We hope to re-frame the understanding of how technology can support the existing creative context [17]. This paper represents the initial attempts to engage with chefs and imagine some ways in which robots could become useful partners in the kitchen. We take inspiration from co-design workshops in which researchers provided farmers experiences around robotic technologies so that the farmers would have deeper insights into the technical possibilities of robots and more quickly imagine a future with robotic tools in the field [20]. In a similar way, we wanted to educate the chefs on robotic technologies and capabilities, but we needed to respect the limited time the chefs had to work with us. We conducted interviews with Danish chefs formally trained in the French tradition of cooking, observed their kitchens to gain insights into their ways of working. Themes that emerged from the initial investigation resonated with some of the elements from fields of art, architecture and theatre and helped to guide the design of nine video prototypes involving the creation and manipulation of food with an ABB IRB120 robot arm [9]. We focused on how the interaction between chef and robot could augment and enhance the creative process. We included examples that highlight unique capabilities of robots and computational support. We showed the video prototypes to ten chefs and waiters and asked for their feedback through a questionnaire and invited them to imagine ways robots could be designed to support them in their work and in the food industry in general.

4.1 Observations

In order to gain a deeper understanding of restaurant kitchens and the work routines within, we conducted three observation and interview sessions. One at a hotel restaurant and the other two at a team-cooking kitchen. Through these sessions, we identified different work processes, goals and agendas unique to the respective kitchens, one focusing more on serving for public diners and the other focuses on the collaborative cooking process. In the observations, we used jotted notes and although we were permitted to ask questions during the observation, we kept these to a minimum and instead reviewed questions and obtained clarifying details from the head chefs after each session during the semi-structured interviews.

The hotel restaurant kitchen was observed at noon, during preparations for the evening and serving à la carte lunch. The head chef's staff consisted of nine chefs with specific roles ranging from grill chef, vegetable chef to pastry chef and various chefs who fill in as needed (roundsman). However, as the staff was urgently preparing ingredients for the night's dishes, more of the staff assisted where needed and took on the role of a roundsman. The chefs were positioned at different stations, e.g. one chef was preparing scallions by peeling them, another prepared meat. The communication was limited to prosaic conversations and small discussions about the ingredients and preparation. The shared knowledge of what each chef was doing predominated the work flow, as one chef would only come over to assist another chef if requested or if certain that he or she needed a extra pair of hands. During lunch, the plating was primarily done in advance with cold dishes or salads being plated at the time of order. Plating was done in layers, built in a bottom-up approach, see Fig. 1.

Fig. 1. The bottom-up arrangement of ingredients in a salad

We developed a sense for how a dish is designed through the discussions with the chefs. The composition and visual expression of a dish starts as an iterative, experimental design process, where the chefs try out different placement, patterns, and arrangement of colors. The plating can either have a specific layout, such as a seasonal color palette or more exploratory with unusual ingredients and varied textures. After an iterative process, the composition of color, tastes and texture manifests in the plating of a dish that subsequently prepared dishes should replicate. This means that a large part of the creative process in cooking, is not a continually on-going process, but rather much of the look and placement of elements is set in the initial stage of creation. Time is a considerable constraint in gastro kitchens and preparation is vital to delivering a high quality experience consistently for all guests. The chefs have to be resourceful and therefore do not experiment or try to be overly creative during busy service hours. Line cooking is predominate in busy kitchens, but the aesthetic composition of the food experience has been meticulously planned beforehand - the diners' experience of a dish is paramount. This is exemplified in Fig. 2 in which the plating has been planned beforehand and then created as copies as consistently as possible.

Fig. 2. In "plating" the chef creates the desired look for an element and then replicates it for each serving

When plating, the chef essentially creates a packaged, edible experience that unfolds as the diners take a bite. After being served, the dish is often explained by the waiter, ensuring that the diners gain an appreciation for the ingredients and the composition - the chef's thought behind the dish. The unfolding experience is a combination of tastes, textures, visual elements and sounds.

In our semi-structured interviews, we asked how they could imagine a robotic collaborator in their kitchen. The answers were focused on the more tedious parts of the job, such as repetitive work and less fulfilling tasks, e.g. when preparing ingredients or cleaning the kitchen. The chefs did not immediately consider how robots could assist in the creative aspects of exploring form and innovative plates. They did, however mention popular consumer robots, such as the iRobot Roomba [5] and the Dyson 360 Eye Robot [4]. They focused on how robots could alleviate the burden of tedious tasks of cutting ingredients, cleaning utensils and other duties that seemed to indicate they imagined robots as "automatons", well suited for executing tasks repeatedly, efficiently and precisely.

In the team training kitchen, the staff consisted of one head chef and two assistants and was observed in the evening. In one session, the head chef focused on teaching a dish and its variations to amateur chefs, who had little experience in a professional kitchen. In another session, the kitchen held team-building exercises by grouping a set of colleagues and having them work together within the kitchen with the final dish creation being a shared responsibility. In both sessions breakdowns occurred within the team due to unclear communication among the team and limited cooking knowledge, and in some cases the participants could not understand the directions given by the expert level recipe. The work was delegated and divided into sub-activities, such as cutting asparagus or opening mussels, where each person had a responsibility for a part of the final dish. The head chef's role was to inform, teach and create a cozy, helpful atmosphere. Plating was done in an ad-hoc fashion whereas the participants would simply experiment with the random placement of ingredients.

Even though the context of the two kitchens was very different, they had some similarities in the way they communicate, divide and delegate work. In both kitchens there is a strong focus on collaborating by dividing the dish up into the preparation and cooking of specific ingredients, where each chef/participant has an area of responsibility. The professional chef replicates one chosen plating of a dish, whereas the team-building participants sought to experiment in a more ad-hoc fashion.

We recognized that issues important to chefs relating to the intended diner experience and means of expression resonate with the concerns of artists, actors and architects. Instead of relegating the robot to copying prior designs, we started to recognize opportunities for the robot to serve as a creative tool for the chef. As an initial exploration, we limit this paper to the exploration and development of design experiences with inspiration from these guiding elements.

4.2   Exploratory Prototypes

We sought out to explore robots providing a role in the creation of aesthetic interactions and experiences regarding the preparation, serving and consumption of food. We developed 9 prototypes serving as an initial exploration into the domain of gastronomy and robotics and are intended to be used to stage further discussions with chefs. These examples aim to help them understand more about the capabilities of robots, how they might manipulate food and support the dining experience. This approach of building initial experiences with new technologies and subject matter experts has been helpful in facilitating communication among the design team to support collaboration in the idea generation process [21].

Taking departure in themes from art, architecture, theatre, and insights gained from the observations and interviews, we developed 9 exploratory prototypes that utilize a desktop industrial robot to prepare food and documented these in short videos. The prototypes were created and documented in the Robot-Lab at the Aarhus School of Architecture over a period of two weeks. In this section we describe the robot platform and each of the prototypes. A compiled short video provides an overview of each of the nine design prototypes [29].

The ABB IRB120 desktop industrial robot [10] as shown in Fig. 3 was used as an experimental platform to support the design explorations. End effectors, tools, or "grippers" are typically connected to the end of the robot arm and are chosen to fit the task. In our case, however, the robot does not come from the manufacturer with specialized food handling effectors, therefore, we designed special purpose grippers and handlers using the Rhinoceros CAD application and then printed them in nylon using the EOS Formiga P110 3D printer [22]. The handling of food and recepticals such as bottles and cooking utensils required careful design work and some quick prototyping with tape, foam, and glue. Some of these custom effectors are shown in Fig. 4. The effectors were designed specifically to the task it should fulfill for example, the spatula-like gripper which can reach under food and move it, the cube-gripper designed for food-cubes that are $3 \times 3 \times 3$ cm in size, etc. Some of the grippers have been modified after being printed to optimize them for the task. The candle gripper was actually the cube gripper, however, by attaching a bit of foam to each gripper, they could grab small birthday candles firmly without crushing them.

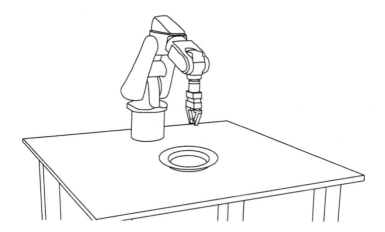

Fig. 3. ABB IRB120 desktop robot mounted to a mobile work surface.

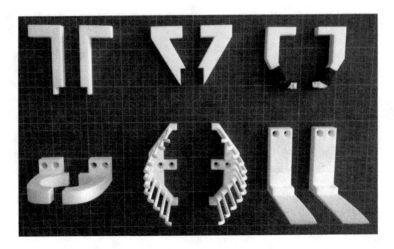

Fig. 4. Effectors supporting the design experiments include various grippers for handling food and food containers (1 cm scale grid)

Our process of creating and conducting the design prototypes, consisted of transforming the conceptual ideas or sketch into a specific task and series of movements, manually programming the movement of the robot and then executing compiled code on the IRB120 robot. The series of movements was created by constructing three dimensional shape of a path in a CAD program, which was then converted to a series of sequential targets with I/O signals and in between positions for smoothness of motion. These movements were then compiled to RAPID code in ABB's RobotStudio [3] and loaded to the IRB120's IRC5 controller [2] to be executed by the robot arm.

Candle. The robot is able to perform repetitive tasks with a high level of precision that creates an opposite aesthetic expression of normal dish, where the organic curves often predominate the plating. It can be seen as the culinary answer to military parades' robot-like marching, which is fascinating, simply because it seems like they posses machine-like precision. In the experiment of Fig. 5; Candle, we draw on the concept of Haute Cuisine, where the robot, carefully and with high precision, places birthday candles at varying angles along a surface and varying patterns.

Plating. Plating is an artistic process in cooking, where the chef creates the foundation for a dish, which complements the main ingredients in both color and shape. Plating is often done in full control of the chef, however we divide the control between robot and chef. In the particular experiment of Fig. 5; Plating 1, the robot creates the boundaries in which the chef can work and forces him to be creative in ways he cannot fully control himself. This is particularly interesting as it challenges the role of the headchef who normally would be plating and be in total control of how the dish is being formed. However, we don't want to

Fig. 5. On the left - Candle: Inspired by Haute Cuisine, the robot picks and places with speed and precision candles in various angles on a birthday cake. On the right - Plating 1: Here the robot creates the boundaries in which the chef has to work with the plating.

replace the headchef and make him, to some extent, obsolete, but we aim to create new ways of how two vastly different entities can collaborate and reach new heights of creativity.

In addition to Fig. 5; Plating 1, we shifted the control between the robot and the chef of how the plating should be conducted. In the example of Fig. 6; Plating 2, it is the chef who creates and decides the boundaries of the plating, wherein the robot has to operate. This is also in striking contrast to what chefs are normally used to, where they are particularly aware of how the end result should be. This is further noted during the observation at the Comwell Hotel as the Head chef stated that the very first plate to leave the kitchen, is the one to copy. Thus, by taking away their control of one of the most important aspect of service, we draw attention to how robots are to be viewed.

Painting. In Fig. 6; Painting, we are taking inspiration in the near surroundings as input and using food as output in such a detailed manner that only artists

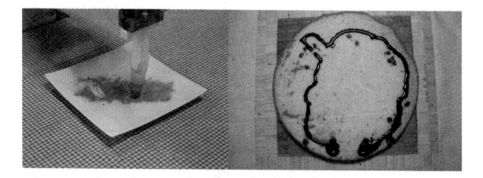

Fig. 6. On the left - Plating 2: The chef creates the boundaries in which the robot can work. On the right - Painting: The robot takes input from the near surroundings and uses food as output to illustrate it.

can perform. In this example The painting of a face or similar can also be seen as a single, exciting experience that takes place in front of the diner - first creating confusion and afterwards surprise, after the diners participate in guessing what the robot is painting. This further enables the chef to form a more personal food experience that can unfold at the table.

Food Visuals. The chef's composition of a dish is often a fusion of the ingredients' taste and color - the combination of taste and visual expression define a cultural dish, like e.g. the overall yellow color palette of an Indian curry dish. In the experiment of Fig. 7; Food Visuals, we investigate how the robots precision and repeatability of actions can create intricate mixes and shapes of colors in food. We investigate how food coloring and geometric shapes can help the chef in discovering new ways to design dishes. The robot can either do a pre-programmed shape in e.g. whipped cream or do random shapes, whilst the chef can add different colors that outlines the shape. In the food coloring experiment, the robot becomes a tool for ideation, which helps the chef compose different colors, shapes and patterns utilizing the precision and randomization of the robot. The chef can simply let the robot control some of the parameters, whilst he maintains the ability to choose the remaining parameters, such as the colors used. This creates an interesting collaboration between robot and chef, as control is negotiated in the beginning. This essentially lets the chef use the robot as a way to investigate different designs, as the robot can be static in its' movements - being the constant, whereas the chef can test different methods or variables.

Fig. 7. On the left: The robot and chef collaborate and experiment with Food Visuals. On the right: The robot works with Modular Cubes from culturally different dished and forms new combinations at the diner's table

Modular Food Cubes. This experiment is tied to the field of architecture. However, the purpose of the experiment is to compose culturally different dishes by a robot using ingredients in cubes. Thereby creating a modular system, and somewhat larger mechanical structure, where the diner's dishes are to be created. The chef creates and prepares the modular ingredients which the robot either systematically (e.g. based on traditional cultural dishes), and randomly, puts together and compose it into a dish at the table.

Fig. 8. On the left - Brownie Wall: Taking inspiration from the field of architecture, the robot builds complex structures from food blocks. On the right - Tension: Tension is built as the robot constructs a structure from food blocks and then destroys it to the surprise of the diner.

Brownie Wall. This experiment, Fig. 8, builds on the field of architecture and how to fabricate complex and modular structures. In this case, the chef prepares the ingredients as modular building blocks for the robot to construct structures for serving. With the precision and accuracy of the robot, the chef will be able to create complex structures with the food thus enhance the food experience of the diners - it can even be built at the table. One could imagine how the robot could continuously build complex structures for each course at a restaurant, letting the diners experience cultural rich food whilst watching the next course being built in front of them.

Tension. This experiment is about building tension and playing on the expectation of the diner to what is going to happen, or likely to happen, in the plating and serving of food. The robot compose a dish and builds it in a seemingly predictable composition. However, as the final details are being made, the entire structure is suddenly contradicted and ruined in a series of rapid and random movements by, the otherwise very precise and reliable, robot. The serving of food is often not tied to meaningful and aesthetic performances as such. The aesthetically pleasing aspect is created, when the chef is plating the dish, but not much change when it is served. Surprises often come down to people not knowing exactly what they are getting or expecting something, but getting something very different.

We want to further play on the serving of a dish and how you can move the expectations of how a dish is composed and served. By letting the robot create a structure that seems to be predictable, to then spoil it at the end, we seek to surprise and impact the dining experience as a whole.

Chaos. Giving complete control to the robot is often tied to the notion of assembly line manufacturing, as the robot is tasked to do same sequence of movements repeatedly. However, in this experiment, we give the robot the control of choosing

Fig. 9. The experiment of chaos. The robot decides what to be served based on a variety of ingredients

ingredients. It can either be chosen through randomization within boundaries, e.g. choosing one type of meat, two types of vegetables or full randomization, simply choosing arbitrary number of ingredients for a dish. Choice can even be based on external sources, such as atmospheric noise, the sound level in the restaurant etc. By randomizing the choice of ingredients, the control shifts away from the chef, and in some cases even the robot, in order to base the construction of a dish on a non-intentional design approach. This can also be seen as a way to surprise guests, sitting together at the same table and ordering different dishes, which the robot can mix together based on known principles, e.g. beef goes well with root vegetables (Fig. 9).

4.3   Reactions, Feedback and Ideation from Prototypes

The exploratory prototypes were created and documented with video, which was then provided for chefs and serving staff to review. A compiled short video provides an overview of each of the nine design prototypes [29]. A questionnaire accompanied the video and was shared on several public forums for Danish chefs and shared directly with two American chefs with a total of 10 respondents. The cross section of respondents was selected to gather feedback from chefs from the Western European perspective, however, we do not claim for this to be an exhaustive inquiry. Rather, we aimed for initial insights and feedback on the prototypes so that we could engage in further discussion about the possible future of the kitchen with robotic agents and identify possible refinements as we continue to develop prototypes for our future design work.

The questionnaire was comprised of a still photo of each prototype from the accompanying video. There were 18 questions, two for each prototype, following the form, *"What was most interesting about this prototype, and why?"*, and *"What aspects of the prototype did you dislike, and why?"* Respondents were assured that there are no right or wrong answers and that our aim was to gain feedback based on their opinions, impressions and that we welcomed any and all feedback they wished to share.

We then discussed the videos in contextual interviews with four of the respondents to gain deeper insights and to probe them for additional feedback and ideas. In the interviews, we explored the beliefs and opinions about robots and then looked for scenarios in which the chef seemed to provide contradictory statements. As described in [32] we wanted to uncover *"...how the subject is solving problems."* We did not seek to confuse or challenge the opinions of the chefs, but rather to explore the mechanics behind their choices and to better understand the conditions in which they accept and embrace robots as a helpful tool as well as when and why they reject them.

The focus of this paper is not to show the most refined prototypes, but to conceptualize the responses and ideas about how future robot-supported food experiences can appeal to the diner and support the creative desires of the chef. We now review the results and insights gained from this process.

5   Emergent Concepts and Ideas

The insights gathered from the design explorations inform our understanding of the design space for aesthetic robot food interaction. Based on the open-ended feedback and interviews in response to the video prototypes, we identify the following key concerns when designing robot-supported interactions with food: issues of control between the human and robot and the perception of robot behaviour.

The chefs were generally supportive of the use of robots - 90 % expressed support for at least some of the prototypes, and 80 % provided key insights about how the individual scenarios can be refined to become more appealing. In the open-ended feedback the chefs expressed an interest in the topic and for more than half of the prototypes. Only one participant provided feedback without specific references to individual prototypes. Some of the respondent chefs did not appreciate our approach in the domain, as they saw it as a direct replacement of him/her even though our proclaimed focus was on collaboration between human and robot - *"Waiters and especially the work of a Chef is craftsmanship - let it stay this way!"*. This resistance to technology was not unexpected, in light of the research that identified a growing trend toward deskilling and attrition due to technology in the food industry [36]. We were delighted, however that most of the chefs opened up and helped to design possible robot experiences for the future kitchen. Other aspects of the prototypes raised concerns for several respondents, e.g. speed and precision, which we will describe more closely in the following sections of the paper.

5.1 Dimensions of Control

Control of ingredient placement has been a central aspect in the development and reflections about the prototypes. From the initial observations in the kitchens, the creation of a dish requires task planning and management of several concurrent processes. However, in the context of human-robot interaction, the current lack of a common language between chef and robot reduces the possibility of negotiation. This forces the chef to rely on the robot as reliable and an active partner instead of a tool. In addition, in order to achieve collaborative control, the chef has to function as a resource that serves the robot, providing information and processing.

In addition, through the language of action, we see examples of the chef creating boundaries for which the robot can work within, Fig. 6. The chef can either give full control to the robot, essentially letting the robot build the dish according to external sources or a pre-programmed repertoire of e.g. patterns and shapes. Fong et al. [23] suggests considering both human and robot needs when designing HRI systems. By giving the robot control, the needs of the robot are central to how the dish is composed, as the robot has to express its needs regarding e.g. ingredients. The chef has to process these needs and react upon them in order to complete the task at hand. The needs can be expressed in explicit and implicit means of gestures. The implicit gestures are categorized as manipulative gestures [12] where it is the actions and motions of the robot that communicates its intentions and needs. Explicit needs are grouped as communicative gestures [12] where pointing and various types of signs are used. This is seen in contrast to how the robot can act as a fully controllable tool, which the chef can choose to use and control just as any other kitchen tool or utensil. This form of control with gesturing either implicitly or explicitly applies to both chef and robot. We elaborate on why non-verbal communication is ideal to use in next section.

In order to give an overview of our exploratory prototypes and how they correlate to the dimensions of control, we have mapped each of them onto a matrix, see Fig. 10. The dimension of control, from Robot to Human, has been visualized along the x-axis of the diagram. Each of the prototypes has been placed according to how they were conducted, however we envision that the majority can be moved to either side of the axis, if control was negotiated differently.

Through the valuable feedback from the chefs and serving staff, additional ideas and concepts emerge as they add key insights to our exploratory prototypes. Even though some people have difficulties imagining robotic agents in the kitchen regarding some of our prototypes, they still seem capable of envisioning how the robotic agents could improve the kitchen and the processes within. An example thereof, is a chef generally being reluctant to robotic agents in the

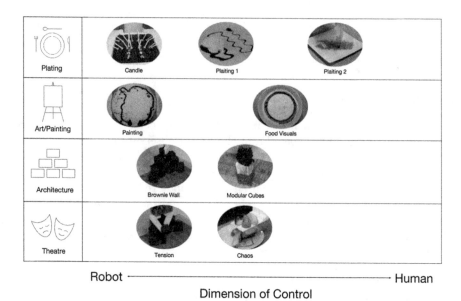

Fig. 10. The experiments mapped to a matrix of categories and dimensions of control

kitchen, who briefly presents an idea to the concept of the "Brownie Wall" prototype. He explains how it would be interesting to introduce the diners with an iPad application in which you build and create structures which the robotic agent in turn converts to actual structures of food by the table in real-time.

In the above example, the role of the chef is the diner along with the control, however, the robot won't acts as a tool in a simple sense, as it becomes an extension of the chef and what the chef/diner is capable of.

5.2   Perception of Robot Behaviour

As a result of our questionnaire, we found that people often perceive robotic movements as either mechanical, i.e. moving one axis at a time in a sequential fashion or more human-like by doing linear movements or moving all axis around a tool-center point - *"It seems artistic in its movements - wavy movements across the plate. But it also resembles a machine when it suddenly stops in the end."* However, as people attribute human-like behaviours to the robotic movements, the robots primary advantages such as speed can have consequences. Thus, seeing a robot complete a task at great speed might resemble that of a person, who does not care for the task-at-hand, thus wants to complete it as fast as possible, regardless of the outcome - *"The robot almost seem as it doesn't wanna do the task. The candles are being placed quick as it just has to be over with..."*. This is seen in contrast to the characteristics of craftsmanship, where the attention to detail is quintessential. In relation to this, much research has already focused on the expressive behaviour of non-humanoid robots and how non-verbal communication [13, 34] can affect the perceived emotional state of the robot based on the robot's motions [39]. As Saerbeck and Bartneck [39] notes, the varying degree of speed/acceleration can have different affects to the perception of the agent conducting a task from being done "carefully" to "aggressively". This corresponds very well to our initial findings from the previously explained prototypes in this section.

Furthermore, when combining both speed and high precision, the robot draws more focus than the object that it is manipulating - the aesthetic experience become more that of amazement of technology and the inner workings of robots than the plating of a dish. In our exploratory design prototypes we find that the robot can either enter the center stage for an experience or be more of a passive actor or tool in the creation of aesthetic experiences. The dimension of passivity correlates with the dimension of control, as the robot takes more control, it is brought to the forefront and the motion is more that of a theatrical performance.

Additionally, when in a collaborative environment, where non-verbal communication is essential and central, perceiving the robot's motions and being able to infer its' intentions and actions is of great importance. In order to gain joint intention the chef and the robot needs to know the intentions of each other [12]. This should be achieved, according to one of the respondent chefs, by gesturing explicitly through communicative gestures to the robot, when asked how she could envision ways of controlling the robot, she answers: "I think the easiest way would... kind of.. grab the arm, physically grab the arm.. and place the arm over?" However, to strengthen the cooperative nature even further, the chef and robot can react on more implicit communication, where the intention of one partner lies within its' actions. By using manipulative gestures [12], verbal communication can be eliminated as it can negatively influence time and efficiency.

Further more, the basic notion of speed and precision is still a key concern as chefs still strive for speed in a kitchen. Tasks have to be finished as quickly as possible. This is a reappearing concern in our online questionnaire, as seen by the responses "With more speed and precision it could work" and "It's too slow, food will be cold before the plate is served", which was in relation to the question - "What aspect of the experiment did you not like?". The attributes of the robot used in the exploratory prototypes was seen as both positive, but also negative as speed, sound and the industrial appearance could affect not only the chef's user-experience, but also the diner's experience when the robot is being placed in the forefront of the dining experience.

In addition to this, in modern society, technology is ubiquitous and we tend to forget about it as it becomes more pervasive. The perception of technology is also at the point that if it does not work correctly, it takes our focus; we get irritated and frustrated. We tend to become oblivious to the complexities and intricacies of the technology that controls the robot. During the realization of the experiments, the authors hypothesised that the diner could become fascinated of the robots' accurate and rapid movement, in the same way as designers and architects embrace complexity in their designs as a way of engaging the viewer [11]. From building a curved wall to careful delivery of a dish, the perception of robots shift in a positive way. In the same way, as a viewer might appreciate a complex structure, he or she could also appreciate seeing how such a powerful, complex robot can be so delicate and precise in a way that surpass human abilities and precision. The appreciation of a robot's movements can be put in the perspective of a theatrical performance as all the robots' axis works both independently and in relation to each other, synchronized and often in a harmonious fashion. The appreciation of complexity is closely tied to how attributes such as sound, speed and precision are perceived by the viewer. The sound of the motors operating within the robot combined with the accelerating and organic motion contributes to the users' experience of the robot.

Consequently, the experiences that unfolds over time are not bounded to the action and task-at-hand of the robot, but can be tied to how it performs these tasks and actions. The diversity of the movements and how it operates while doing an otherwise dull and repetitive task, forms new experiences for the spectator as it differs from the norms. This could be further emphasized by adapting the method of Saerbeck et al. and their use of the PANAS and SAM scales for assessing affect in relation to motion of the robot [39].

6 Discussion

We outline key concerns for aesthetic food interactions supported by robots that we propose can be useful for making sense of the design space and opportunities for exploration with future work. In addition to the design of technology supported experiences, we contribute to an understanding of how people experience food, which has been an activity that has involved tools, ritual and cultural influences well before digital technologies entered the stage. In addition, we broaden the discussion of how we perceive robots as an entity we collaborate and interact with to create an aesthetic food experience. Hence, robots are not merely a tool to obtain efficiency, but can be enriching in a collaborative environment as is the case with the modern kitchen.

6.1   Robots in the Forefront of Food Experiences

The roles of robots are rather firmly rooted in the existing examples of service robots, industrial manufacturing, etc. Placing the robot in forefront of the experience of a diner suggests further scenarios to be explored. As noted earlier, letting the robot perform its' tasks in the view of the diners seems to create varying degrees of aesthetic experiences. People want to explain what is happening and struggle to make sense of its movements. In most cases the chefs explain the behavior of the robot as if it is a human being. This anthropomorphization includes ascribing human-like intention and perceived personalities, which means they tend to treat these types of machines as social entities. Furthermore, the careful movements as a waiter noted, are often explained as being intriguing or mesmerizing. This probes some interesting questions of the robot as a social actor and what role it has along with social skills [19].

Throughout the design explorations, the robot can be positioned in various stages of a dinner experience. The role will then depend on the perspective of the person looking at it. A chef can see it as a partner/companion or a simple machine/tool according to the work of [19], to enhance his own creative process, while the diner can see it as a chef, waiter or even a social agent as part of the actual experience of eating.

In the prototype demonstrating "Tension", Fig. 8, the robot builds a structure, which the diner expects to be completed, only to find that the robot destroys what it had been building, thus sparking feelings of asperity. Why would it destroy something it had spent time building? It suggests that the chef might not be in complete control of the robot causing a sense of apprehension for the diner. In addition to this, the robot takes on a role by itself as a chef or waiter putting itself at the center of the dining experience. The purpose of this is not to destroy dishes and frustrate diners, but merely a way to entertain and surprise diners at the table. In many dishes, we have certain expectations to how it is prepared and presented, this preconception can be challenged directly in front of the diner.

Furthermore, during the "Painting" prototype, viewers might embrace the personal attribution it imposes on the dish. Thus, appreciating the presence of the robot and how it contributes to the social experience. This can be further exemplified in the prototype exploring, "Chaos", where the diners rely on the robot to serve a dish to their likes. It pushes the limits of the dining experience as the norm prescribes that you get what you ordered. However in this example, control is partly given to the robot, as it is the entity to ultimately decide what to serve based on the chef's prepared ingredients from the diner's original order.

Depending on where the robot is placed, we see contradicting statements regarding what role the robot should adopt. One chef noted in general, that the robot should only be used as a tool or extension of the chef, but never replace the chef. However, the particular respondent had no issues with delegating some of the human waiter's tasks to a robot: *"What do you call it.. Saving money on waiters, so they [the robots] become the waiters and the setup... Kind of... Go down and light up candles, give a presentation of the menu, while the [human] waiter is pouring wine."* These contradicting statements illustrates some of the fears that the chefs have regarding the use of robots in the gastronomical world.

6.2 Robots in the Background as an Active Partner for the Chef

As stated earlier in our research problem, we seek to give the robots more substantial roles in the kitchen alongside chefs and serving staff. Fulfilling in a way, where the robot contributes to the creation of culture, not just taking over laborious tasks that are seen as constraints for the chef's creativity, but instead taking part and contributing to this creativity more directly.

When working in conjunction with chefs, the number of design parameters can be increased by using the robot as a fabrication tool. An example of this can be seen in our food coloring prototype, where the robot can stir in complex patterns.

If we further develop this line of thinking, the observed chefs noted robotic agents could be beneficial in repetitive tasks, but did not want to spend much time in instructing the robots. Perhaps for cutting vegetables, a chef could direct the robot to cut in specific motions and patterns, relying on the kinaesthetic experience of interacting with the kitchen tools in their hands to signal and control the robot. This correlates well with the existing praxis of communication through action and hand movements that takes place within the kitchen staff. Instead of supporting existing practises in kitchens, a robot could also seek to alter them, such as the process of designing a dish. An example could be that a robot working within a range of options to plate and continuously change the plating over time as opposed to the more static process of designing and afterwards, replicating. As seen in our observations, the actions of a chef's cooking can also communicate needs to the near surroundings, which causes spontaneous collaboration and assistance between chefs. In a similar fashion, a robot could take a non-intrusive role of an assistant or even operate as an extension of the chef. An example of this could be a Chef plating two of the same dish, whereas the robot would replicate the design of the dish that the Chef is currently plating. The robot and chef could also work in shifts when plating, each placing an ingredient in relation to what has just been placed, such as seen with our plating experiments. By using the ingredients as means for communication, such as the plating with chocolate powder experiment, the communication happens through simple gestures that are contextual and explicit.

By focusing on the robot as a mentor instead of simply a collaborator or assistant, the robot could also take on the role as a scaffolding tool, helping Range chefs learn the specific tasks and routines of a cooking station or simply new plating designs or techniques.

So far we have discussed the robot as being visible and integrated to the experience, however, the robot can be helpful in various other ways. For example, the robot can act as a dynamic jig for the placement of objects. The robot could also be used as a creativity toolkit that helps Chefs explore and develop new plates that they then can do later by hand. There are many ways to imagine robots taking up roles alongside and supporting Chefs and serving staff–our explorations have only begun to open the design space.

7   Conclusion and Future Work

We have, through design explorations, seen examples of how the chef and robot can collaborate to create experiences for diners and even for the chef himself. Using a robot in various methods of cooking can enhance creativity as new possibilities of methods in handling food opens up.

The exploratory prototypes developed in this project have only sketched some of the directions in which a robot and chef can collaborate. The kitchen is one of the most important culturally significant contexts that shapes our everyday lives. Old traditions are taught from parent to child and these traditions move around with us in a globalized world. One can easily catch a glimpse of other cultures, simply by dining out at one of the many restaurants found in the modern city. In our explorations, we have focused mostly on western food, specifically Scandinavian cuisine, however traditions from other cultures should be investigated and might provide additional insights into how a chef might benefit from working a robotic collaborator. By investigating the food related traditions from other cultures and through engaging with chefs, serving staff, and diners from a wider perspective, we expect to uncover additional and divergent responses towards robotic assistants in the kitchen. This continued exploration into other cultures marks exciting and important directions for future research.

The exploration of control between robot and chef can be further investigated by implementing it into existing processes or by creating entirely new processes. Current processes dictate that the visual expression of dish is static after an ideation phase, which the robot could support by taking and giving control to the chef, forcing her to investigate plating through a more unexpected experimental process. However, we could also imagine that plating was not only limited to the creation of a dish, but instead redesigning the dish continuously each time it was served. As seen in our experiment with randomization, the dishes could also be plated uniquely in front of the diners, creating a more personal experience and possibly greater appreciation for the food and the experience surrounding the consumption of food.

A similar future direction for the dimension of control is to investigate how the level of control correlates to creativity. Relating to Csikszentmihalyi's concept of Flow in positive psychology [18], it would be interesting to explore how the robot could balance and adjust its contributions to maintain the challenges presented to the chef, thus maintaining a state of flow. This and other opportunities made possible by robots working alongside humans signal important work.

Lastly, in relation to the flow of creativity, the language between human and robot is an interesting topic for further investigation. In the "Plating 1", Fig. 5, the chef or robot creates a boundary through manipulation of an ingredient, thus communicating intention through action. We invite exploration into verbal as well as non-verbal communication for these situations of real-time coordination. Cooking involves manipulation of physical ingredients and tools and we hope that our work inspires new experiments into the domain of contextual, gestural/action-oriented communication.

References

1. ABB Food and beverages: http://new.abb.com/food-beverage. Accessed: 12 Aug 2015
2. ABB IRC5 Controller: http://new.abb.com/products/robotics/controllers/irc5. Accessed: 10 Aug 2015
3. ABB RobotStudio: http://new.abb.com/products/robotics/robotstudio. Accessed: 10 Aug 2015
4. Dyson 360 Eye: https://www.dyson360eye.com/. Accessed: 12 Aug 2015

5. iRobot Roomba: http://www.irobot.com/For-the-Home/Vacuum-Cleaning/Roomba.aspx. Accessed: 12 Aug 2015

6. Sorter pro packing robot: http://www.sorter.eu/en/sortingmachines1/pro-packing-robot,1315.html. Accessed: 14 Aug 2015

7. Universal Robot Food and agriculture: http://www.universal-robots.com/industries/food-and-agriculture/. Accessed: 12 Aug 2015

8. TNS opinion & social: Public attitudes towards robots (2012). http://ec.europa.eu/public_opinion/archives/ebs/ebs_382_sum_en.pdf

9. ABB: IRB 120 - industrial robots (2015). http://new.abb.com/products/robotics/industrial-robots/irb-120

10. Abb: IRB 120 - industrial robots (robotics) - industrial robots - robotics - ABB, June 2015. http://new.abb.com/products/robotics/industrial-robots/irb-120

11. Alexiou, K., Johnson, J., Zamenopoulos, T. (eds.): Embracing Complexity in Design, 1st edn. Routledge, October 2009. http://www.worldcat.org/isbn/0415497000

12. Bauer, A., Wollherr, D., Buss, M.: Human-robot collaboration: a survey. Int. J. Humanoid Robot. 5(01), 47–66 (2008)

13. Breazeal, C., Kidd, C., Thomaz, A., Hoffman, G., Berlin, M.: Effects of nonverbal communication on efficiency and robustness in human-robot teamwork. In: 2005 IEEE/RSJ International Conference on Intelligent Robots and Systems, IROS 2005, pp. 708–713, August 2005

14. Cahlander, R., Carroll, D., Hanson, R., Hollingsworth, A., Reinertsen, J.: Food preparation robot, 1 May 1990. https://www.google.com/patents/US4922435, uS Patent 4,922,435

15. Chang, W.L., Šabanovic, S., Huber, L.: Use of seal-like robot PARO in sensory group therapy for older adults with dementia. In: Proceedings of the 8th ACM/IEEE International Conference on Human-robot Interaction, HRI 2013, pp. 101–102. IEEE Press, Piscataway (2013). http://portal.acm.org/citation.cfm?id=2447587

16. Comber, R., Choi, J.H., Hoonhout, J., O'Hara, K.: Designing for human-food interaction: an introduction to the special issue on 'food and interaction design'. Int. J. Hum. Comput. Stud. 72(2), 181–184 (2014). http://dx.doi.org/10.1016/j.ijhcs.2013.09.001

17. Cross, N.: Designerly Ways of Knowing, 2nd edn. Birkhauser Verlag AG, April 2007. http://www.amazon.com/exec/obidos/redirect?tag=citeulike07-20&path=ASIN/1846283000

18. Csikszentmihalyi, M., Csikzentmihaly, M.: Flow: The Psychology of Optimal Experience, vol. 41. Harper Perennial, New York (1991)

19. Dautenhahn, K.: Socially intelligent robots: dimensions of human-robot interaction. In: Emery, N., Clayton, N., Frith, C. (eds.) Social Intelligence: From Brain to Culture. OUP Oxford (2007)

20. DiSalvo, C., Fries, L., Lodato, T., Schechter, B., Barnwell, T.: GrowBot garden (2010). http://carldisalvo.com/posts/growbot-garden/

21. DiSalvo, C., Lukens, J., Lodato, T., Jenkins, T., Kim, T.: Making public things: how HCI design can express matters of concern. In: Proceedings of the SIGCHI Conference on Human Factors in Computing Systems, CHI 2014, pp. 2397–2406. ACM, New York (2014). http://dx.doi.org/10.1145/2556288.2557359

22. EOS: FORMIGA p 110 - laser sintering 3D printer for rapid prototyping - EOS, June 2015. http://www.eos.info/systems_solutions/plastic/systems_equipment/formiga_p_110

23. Fong, T., Thorpe, C., Baur, C.: Collaboration, Dialogue, Human-Robot Interaction. In: Jarvis, R., Zelinsky, A. (eds.) Robotics Research, Springer Tracts in Advanced Robotics, vol. 6, pp. 255–266. Springer, Heidelberg (2003). http://dx.doi.org/10.1007/3-540-36460-9_17

24. Foster, M.E., Keizer, S., Lemon, O.: Towards action selection under uncertainty for a socially aware robot bartender. In: Proceedings of the 2014 ACM/IEEE International Conference on Human-robot Interaction, HRI 2014, pp. 158–159. ACM, New York (2014). http://dx.doi.org/10.1145/2559636.2559805

25. Fukuda, T., Nakauchi, Y., Noguchi, K., Matsubara, T.: Human behavior recognition for cooking support robot. In: 13th IEEE International Workshop on Robot and Human Interactive Communication, ROMAN 2004, pp. 359–364. IEEE (2004). http://dx.doi.org/10.1109/roman.2004.1374787

26. Gaver, W.: What should we expect from research through design? In: Proceedings of the SIGCHI Conference on Human Factors in Computing Systems, CHI 2012, pp. 937–946. ACM, New York (2012). http://dx.doi.org/10.1145/2207676.2208538

27. Gravot, F., Haneda, A., Okada, K., Inaba, M.: Cooking for humanoid robot, a task that needs symbolic and geometric reasonings. In: Proceedings 2006 IEEE International Conference on Robotics and Automation, ICRA 2006, pp. 462–467. IEEE, May 2006. http://dx.doi.org/10.1109/robot.2006.1641754

28. Grimes, A., Harper, R.: Celebratory technology: new directions for food research in HCI. In: Proceedings of the SIGCHI Conference on Human Factors in Computing Systems, CHI 2008, pp. 467–476. ACM, New York (2008). http://dx.doi.org/10.1145/1357054.1357130

29. Laursen, C.Ø., Pedersen, S., Merritt, T.R.: Robot-supported food experiences: exploring aesthetic plating through design experiments, June 2015. https://vimeo.com/130385289

30. Lee, H.R., Sung, J., Šabanović, S., Han, J.: Cultural design of domestic robots: a study of user expectations in korea and the united states. In: 2012 IEEE RO-MAN, pp. 803–808. IEEE (2012). http://dx.doi.org/10.1109/roman.2012.6343850

31. Ma, W.T., Yan, W.X., Fu, Z., Zhao, Y.Z.: A chinese cooking robot for elderly and disabled people. Robotica 29(6), 843–852 (2011). http://dx.doi.org/10.1017/s0263574711000051

32. Mitchell, A., McGee, K.: Limits of rereadability in procedural interactive stories. In: Proceedings of the SIGCHI Conference on Human Factors in Computing Systems, CHI 2011, pp. 1939–1948. ACM, New York (2011). http://dx.doi.org/10.1145/1978942.1979223

33. Robotics, M.: Robotic kitchen. http://www.moley.com/

34. Novikova, J., Ren, G., Watts, L.: It's not the way you look, it's how you move: validating a general scheme for robot affective behaviour. In: Abascal, J., Barbosa, S., Fetter, M., Gross, T., Palanque, P., Winckler, M. (eds.) INTERACT 2015. LNCS, vol. 9298, pp. 239–258. Springer, Heidelberg (2015)

35. Palep, J.H.: Robotic assisted minimally invasive surgery. J. Minimal Access Surg. 5(1), 1–7 (2009). http://dx.doi.org/10.4103/0972-9941.51313

36. Robinson, R.N.S., Barron, P.E.: Developing a framework for understanding the impact of deskilling and standardisation on the turnover and attrition of chefs. Int. J. Hospitality Manage. 26(4), 913–926 (2007). http://dx.doi.org/10.1016/j.ijhm.2006.10.002

37. RS TECH: Robotics Automation - Projects, 2. http://www.rstech.com/projects.html. Posted-at 12 Oct 2015

38. Šabanović, S., Bennett, C.C., Lee, H.R.: Towards culturally robust robots: a critical social perspective on robotics and culture. In: Proceedings of HRI Workshop on Culture-Aware Robotics 2014 (2014)
39. Saerbeck, M., Bartneck, C.: Attribution of affect to robot motion. In: 5th ACM/IEEE International Conference on Human-Robot Interaction (HRI 2010), pp. 53–60. ACM, Osaka (2010)
40. Samani, H., Saadatian, E., Pang, N., Polydorou, D., Newton Fernando, O.N., Nakatsu, R., Valino Koh, J.T.: Cultural robotics: the culture of robotics and robotics in culture. Int. J. Adv. Robotic Syst., 1+ (2013). http://dx.doi.org/10.5772/57260
41. Sanders, E.B.N., Stappers, P.J.: Co-creation and the new landscapes of design. CoDesign 4(1), 5–18 (2008). http://dx.doi.org/10.1080/15710880701875068
42. Shah, J., Wiken, J., Williams, B., Breazeal, C.: Improved human-robot team performance using chaski, a human-inspired plan execution system. In: Proceedings of the 6th International Conference on Human-robot Interaction, HRI 2011, pp. 29–36. ACM, New York (2011). http://dx.doi.org/10.1145/1957656.1957668
43. Shi, J., Jimmerson, G., Pearson, T., Menassa, R.: Levels of human and robot collaboration for automotive manufacturing. In: Proceedings of the Workshop on Performance Metrics for Intelligent Systems, PerMIS 2012, pp. 95–100. ACM, New York (2012). http://dx.doi.org/10.1145/2393091.2393111
44. Steinfeld, A., Fong, T., Kaber, D., Lewis, M., Scholtz, J., Schultz, A., Goodrich, M.: Common metrics for human-robot interaction. In: Proceedings of the 1st ACM SIGCHI/SIGART Conference on Human-robot Interaction, HRI 2006, pp. 33–40. ACM, New York (2006). http://dx.doi.org/10.1145/1121241.1121249
45. Turkle, S.: A nascent robotics culture: new complicities for companionship. Online article (2006)
46. Wallin, P.J.: Robotics in the food industry: an update. Trends Food Sci. Technol. 8(6), 193–198 (1997). http://dx.doi.org/10.1016/s0924-2244(97)01042-x
47. Wei, J., Wang, X., Peiris, R.L., Choi, Y., Martinez, X.R., Tache, R., Koh, J.T.K.V.,, Halupka, V., Cheok, A.D.: CoDine: an interactive multi-sensory system for remote dining. In: Proceedings of the 13th International Conference on Ubiquitous Computing, UbiComp 2011, pp. 21–30. ACM, New York (2011). http://dx.doi.org/10.1145/2030112.2030116
48. Yu, Q., Yuan, C., Fu, Z., Zhao, Y.: An autonomous restaurant service robot with high positioning accuracy. Ind. Robot: Int. J. 39(3), 271–281 (2012). http://dx.doi.org/10.1108/01439911211217107

Robots as Producers of Culture:
Material and Non-material

'Face Robots' Onscreen: *Comfortable and Alive*

Elena Knox[✉]

University of New South Wales, Sydney, Australia
e.knox@unsw.edu.au

Abstract. This chapter draws upon the author's experimental video artwork *Comfortable and Alive*, made with the Japanese gynoid robot Geminoid-F by ATR Hiroshi Ishiguro Laboratories, to facilitate a wider, yet fractional, discussion of the cultural provenance and potential integration of specifically female-appearing android robots (gynoids). The "display architecture" of the gynoid can be viewed as an aesthetic emulation by robot designers of the centuries-old characterization of girls and women as naïve, pretty, submissive and soothing; this construction also pervades televised and other media. At the present time it is viewed as ideal that the service gynoid should make humans feel comfortable, most often in companionship, entertainment, hostessing, and reception roles. The artwork raises poignant issues pertaining to machine translation, and human–machine affinity, in context of the replication in robots of societal gender norms.

Keywords: Gynoid · Gender · Robotic affinity

1 Introduction

The robot Geminoid-F by ATR Hiroshi Ishiguro Laboratories has recently appeared in a feature film, *Sayônara* [1]. But before this, in 2012–14, I worked with the same robot to create a suite of 6 video scenes called *Actroid Series I*, which introduced the Geminoid as a film actor in speaking roles. Some scientists personally involved with Geminoid-F (or Actroid-F, as it is also known) describe this android robot as being 'more believable onscreen' [2], based on their own and their participants' interaction with it during cognitive and affective experiments. As an artist, I have taken up this proposition by presenting the android "participating" in its audiovisual mediatization, via video art, in order to record and scrutinize a technocultural moment in the human pursuit of developing 'very humanlike robots' [3]. *Actroid Series I* offers, to my knowledge, the first screen-based dramatic scenes of an Actroid being verbal; they are certainly the first experimental art of an Actroid orally engaging with its own ideologic construction (see for comparison the simple live experiment [4] with Geminoid-F conducted in the same year as this filming, in which an android, a human and a box read poetry to 17 university students one by one in a room). Sifting through its cultural provenance in human and machine dreams, coded routines, institutional laboratories, and long histories of power and trade, the Actroid in *Comfortable and Alive* speaks in monologue in the tradition of the theatrical "talking head".

© Springer International Publishing Switzerland 2016
J.T.K.V. Koh et al. (Eds.): Cultural Robotics 2015, LNAI 9549, pp. 133–142, 2016.
DOI: 10.1007/978-3-319-42945-8_11

To achieve this articulation of an amplified, self-reflexive android ideology I have focused herein on the Geminoid's face as interface. Claudia Springer wrote that the 'pleasure of the interface, in Lacanian terms, results from the computer's offer to lead us into a microelectronic Imaginary where our bodies are obliterated and our consciousness integrated into the matrix' [5, 37]. So, most of the videos in the *Series* explore and portray the contemporary android as a somewhat clumsily materially embodied but cognitively awake and aware artificial intelligence (the former is more or less true; the latter is a creative projection beyond their current reality, which as telepresence androids relies very much on "wizard-of-Oz" chicanery and veiled human puppeteering). The robot's professed sentience seems to work to transcend its functional bodily limitations. Its small facial performance trajectories not only express but also elicit affect in ways poetic and empathic; this is the job of the actor in the arts.

In addition to tendering a philosophical critique, *Actroid Series I* marks this technocultural moment by cataloguing Actroid-F's replicative behavioural units in tightly-framed and simply-programmed performances across a range of parameters in expressive movement. It exhibits the gynoid smiling, blinking its eyes, and talking (the speech act comprises text programming and sound production in association with mouth movement). It exploits this robot's capacity to roll its eyeballs upward and downward, and to perform a "prepackaged" expression of "surprise" (mouth open, eyebrows raised). The video artworks incorporate computer-based language translation, text-to-speech software, head movement controlled by Geminoid-F's graphical user interface, some arm gestures, and the robot model's remarkably variable facial mask/mirror as a window into the android "soul". This approach chimes with its developers' current focus on the head and face over the largely "placeholder" body [6]. In this short chapter 1 will discuss the ideas behind my 6-channel video installation *Comfortable and Alive*, organized under the themes of translation, bionic affinity, gender, rebellion, and replication.

Androids are sometimes called "face robots". Geminoid-F has a face modelled on an anonymous 20-year-old [7]; the face of her precursor Repliee Q1 was modelled on newscaster Ayako Fujii. Otonaroid and Kodomoroid, developed in 2014, are also molded on real-human faces [8], and their 2015 robot Erica is created as a composite of classic and standardized facial attractiveness. Ishiguro's method for designing Erica's face was as follows:

> The principle of beauty is captured in the average face, so I used images of 30 beautiful women, mixed up their features and used the average for each to design the nose, eyes, and so on... [9]

Given that the research focus is on eliciting human affinity via cranial and facial simulation, Geminoid-F has a surprisingly economical 11 motors in its head. The most advanced Korean humanlike robot, EveR-4 (and EveR-4 M), has 33 motors in its head. The EveR series (for the Biblical Eve + R for robot) are an all-female entertainer and service android group. They 'are developed in particular with a focus on the facial part, because the appearance and expression capability of the face has a significant effect on the use of the robot' [10, 2300].

Moon-hong Baeg, senior researcher at the Korea Institute of Industrial Technology, has conducted ongoing "plastic surgery" on the EveR gynoids [11]. As in the case of the Japanese Actroids, the constant modification of the face and physique of the EveR is framed in terms of making the robots more appealing and comforting to humans, and appears to be motivated by familiar ideals and standards regarding beauty, gender, and corporeality. The process might also be seen as metaphorically recalling a gruesome past in which corporeal experiments and "anatomy theatre" were conducted on societies' less powerful groups [12, 34–58, 13, 52–53, 14]. Concerning the previous EveR models, 'Baeg and his men ... thought EveR-2 was much prettier and more human-like than EveR-1, which looks like a doll', but were yet unsatisfied and 'opted to conduct a facelift on EveR-2 while making its hands smaller' [15]. As is also the case with Actroids [2, 16] the designers are uncomfortable with the realistic size of the hands they have made, as they are not in concert with popular images of women with dainty, feminized, and unrealistically tiny hands and feet. In today's saturated mediascape of modified images implicitly promoting the reduction and restriction of women's most dextrous and useful motor appendages (the idea recalls past practices of female foot-binding), it might be little wonder that the current gynoids' hands, even when molded from actual women, appear "too large" if sized to contain enough motors to properly do their jobs. This is another reason why, at present, the mostly-female Geminoid and EveR ranks are at their most effective from the neck up.

Android building extends and externalizes rationalist principles of itemization, wherein it is posited that '[a]ffective interaction can be achieved through the replication of behaviours understood to comprise it, made up of units assembled into a catalogue of affective expressions, productions, recognitions, and normative responses' [17, 233]. Actroids are described by their creators as conveniently modular – 'You can take my androids on planes – the torso in the suitcase and the head in carry-on', jokes Ishiguro [18]. In literalizing this comically ignoble mind–body split in my videos, I amplify the Cartesian (and arguably anachronistic) "carry-on", as the Actroid catalogues, enumerates, and wittily deconstructs its culturally and technically embedded performative functions.

2   Translation

The monologue performed by Geminoid-F in *Comfortable and Alive* consists of freely downloadable hypnotic induction scripts available on the Internet. The "user" of the artwork can choose between 6 languages: Russian, Chinese, Japanese, Portuguese, English, and Korean. In a creative process that loops around the information super-highway, the texts have been translated from English into the other languages somewhat imperfectly through my use of online software. No attempt has been made to clean up the translations, as this is deemed to be beyond this android's "mental" and communicative capacities and skills at the present time. Plugged into the network and accompanied by computer-generated ambient "new age" music (an original score composed to sound as generic as possible), the Actroid urges the viewer/listener to enter a trance state.

Without thinking about it, you will soon enter a deep, peaceful, hypnotic trance, without any effort ... every nerve, every muscle, every fiber continues to melt ...

Excerpt from *Comfortable and Alive*

By virtue of the statistical machine translations, cultural (in)sensitivities and nuances are deliberately glossed over and transmission is approximated; yet the intention is clear and the aesthetic familiar, due to the generic nature of the ostensibly soothing message, and also to the generic engineering approach to emotive and expressive facial communication (see e.g. [10]). I have shown *Comfortable and Alive* to native English, Japanese, Portuguese, and Chinese speakers. Each group gave feedback that despite the linguistic imperfections resulting from machine translation, the Actroid looked "more natural" (on the video) speaking their native language.

The Japanese Government's 2007 charter *Innovation 25* contains cartoon portrayals of a future robot-dependent society [19, 20]. The illustration for the fifth subsection of this governmental charter is captioned (in translation by Google) 'spread of machine translator, can communicate with people of all nations' [19]. In this illustration, a large man is being served by a slim hostess, presumably on an international flight. The hostess is providing, not just food, drink, amenities and visual stimulation, but also a high-tech means for (the) man to express his needs and desires to be serviced by her. Gender is encoded in her stance (he is sitting, as denotes authority in Japan and elsewhere), and her own desires are presumably sublimated in her embodiment as medium for the translated message. So, as day-to-day speech translation by and through machines becomes more ubiquitous and imperative on the way to the heralded year 2025, what kinds of being does the practice of machine translation *translate to* for different stakeholders in the interaction?

3   Gender

In its "maturity", and even retaining its -F for female, Geminoid-F has become the subject of a mitotic pseudo-sex change. 'AIST has developed a male version of its android robot Actroid-F', Japan's National Institute of Advanced Industrial Science and Technology (AIST) declared in 2011 [6]. Yoshio Matsumoto, of Intelligent Systems at AIST and cross-institutional collaborating colleague of Hiroshi Ishiguro, explains the decision as cited below. (Three minutes into this video interview, the graphical interface used to manipulate the Actroids' movements in real time is displayed onscreen. This interface is very similar to that used in the creation of *Actroid Series I.*)

> We often get asked why we made a female robot, so we decided to make a male version as well. Mechanically, the two robots are the same, and if you take the wigs off, the faces are the same too. ... We think this makes Actroid-F a good test platform for comparing the impressions made by male and female robots [6].

Since the announcement of the "male" version of Actroid-F, ATR Laboratories with AIST have developed the following very humanlike robots: new "female" worker Actroids (Kodomoroid, Otonaroid); the gynoid Erica, which according to Ishiguro 'is the "most beautiful and intelligent" android in the world' [9]; improvement and divergence in the existing geminoid copy of Ishiguro himself (HI-2, HI-4); another copy of a male professor (DK); and YangYang, an adult "female" Geminoid in collaboration with a Chinese company. There has been much touring and exhibition of "female"

Actroid-F and the other "females"; but little if any further publicity for the reactive Actroid-F "brother".

AIST's claim that it developed a male robot is interesting. In a reversal of the Eve-from-Adam myth, the "male" robot was a clone of the "female" one with same face, short wig, different clothes and reshaped foam padding. "She" wore a pink cardigan while "he" wore a blue tie. Overnight, the robots became effigies to be presented to humans as stable-sexed experimental stimuli. The understanding of gendering, and even sexing, robots as a clothing procedure accords with the Butlerian principles of gender performativity [21], but is a shallower and more abrupt interpretation. Though there is for Matsumoto's team no *a priori* difference between these two now gendered bodies, there is also no scope for gradual development and no fluidity between typologies, hence *activity* is missing in the process of active societal gendering (cf. [20, 4–5]). The robots' genders are fabricated, but they are passive and polarized. As the independent variables in a scientific 'test platform' [6] they are, ironically, potentially quite variable but to date have only been considered as variable in binary opposition. Their independence as variables is simplistically conceived as absolute, but they are not independent from the scientists' own gender performativity and concept thereof. If gender is a corporeal technology [22], then the tokenistic Actroid "brother" is technology-drag; this without the subversive kinesis or promise of the drag arts, which find images of flexibility amongst what J. Jack Halberstam terms the 'continued refusal in Western society to admit ambiguously gendered bodies into functional social relations' [23, 15]. The Actroid as drag king (in the sense designated by [20, 22, 23]) *could*, but for this refusal (both Western and Eastern), signal diversity in the prescriptive gendered compound, as a non-standard and multifaceted "identity" product (Fig. 1).

Fig. 1. Elena Knox, *Comfortable and Alive* [video still], 2014, 6-channel installation with alternate 6 soundtracks, HD, 3:22 seamless loop.

4   Gynoid Insurrection

Jeffrey Koh [24] defends implications by some general public respondents to the Singapore-based Lovotics project that Lovotics robots are or can be manipulative, by saying: 'It's not as if robots are seducing people or using some hypnotic methods.' *Comfortable and Alive* plays with this idea, wondering whether these events will in fact be possible in future, and to what extent they incorporate persuasion being dependent on self-seduction, self-hypnosis. The Actroid in character as hypnotist in *Comfortable and Alive* uses her affected social status as commander, however brief, to impress a version of performativity on the hypnotized that will presumably inhere post-hypnosis. The affective state she attempts to impress is a gendered and compromised one. It could be read as profoundly embodied, and conversely as profoundly disembodied, dispersed throughout the textures of the cyberworld.

> *When there is nothing for you to say, you simply glow with internal radiance, showing honesty, sympathy and concern. You are so interested in people that you stimulate them to fully express their views – before you speak. When words are inadequate you act out the emotions you feel; you demonstrate a level of maturity and perception which is rare.*
>
> Excerpt from *Comfortable and Alive*

The mode of sympathetic relation described in the above quote, while a worthy one and by no means undervalued or ridiculed here, is also silent, subjunctive and primarily gestural. The Actroid-hypnotist bestows this role upon another. At the same time as she demonstrates the relational mode's gestural boundaries via mirror-screen, teaching her trance-bound subject by physically acting out the specified pauses, smiles, radiant sympathetic glow and quiet lively concern, Geminoid-F verbalizes her instructions clearly and methodically. The amalgam of cognizant signification and embodied cognition amounts to a *re-programming* of the "other" under hypnosis. This other is "alive" in a sensual sense, presumably bio-human, indeterminate on the sex/gender spectrum and yet definitely above the Actroid in the hierarchy of cyber–organic agency. In an optimistic film/dream, the hypnotizing robot hostess rehearses a rewrite of her role in society. The haptic visuality [17, 279] of this rewrite is achieved through embodied doubling of her head-and-shoulders and a mimetic engagement with her usual disposition.

One of Ishiguro Laboratories' latest gynoids is 'the adolescent-looking "Kodomoroid", an amalgamation of the Japanese word "*kodomo*" (child) and "android"' [18]. A remote-powered machine that does not currently have the artificial intelligence to learn or to make decisions autonomously, it is presented in the press as having dreams and a sense of humour. 'We want robots to become increasingly clever', says Ishiguro [18]. 'My dream is to have my own TV show in the future', Kodomoroid said (in surrogacy) at its press preview [18]. The conceit of the artwork *Comfortable and Alive* is that Actroid-F is aware of her reputation in arts and culture as a brainwashed drone. In her multilingual monologue to camera she turns the tables on this perception, urging her own dominance over the interactive situation, her trustworthiness, her authority, and her situated autonomy. She requests her subject to 'be comfortable'; thereafter she gives orders during the subject's 'deep, peaceful, hypnotic trance'. There is a sinister flavor to this usurpation of agentic power: the Actroid knows, through informatically absorbing science fiction, that robot will-to-power is something many humans imagine and, crucially, fear. The takeover is bracketed by the prompt or hypnotic suggestion that the trance will be 'as automatic as dreaming' and that its events will be forgotten upon awakening. Human becomes automaton [25, 228–230].

> *If I touch you, if I touch you in any way, do not be alarmed, do not be intimidated by my touch,*
> *just let it be your signal to let go and relax and melt down further and further ...*
>
> Excerpt from *Comfortable and Alive*

Anything could happen in this state and it has all happened to hostesses before. In *Comfortable and Alive* Actroid-F mimics the oppressor's rhetorical minimization of harm, whilst also satirizing the ambivalence of humans to being touched *by* her – the fetishist discord between her base reception as concomitant living computer, revulsive corpse and enigmatic sex object [5, 20, 26, 27].

5   Bionic Affinity

With the increasing permeability of boundaries between model and medium, matter and materialization, it seems that we may indeed progress toward a science-culture (sci-cult?) fiction-future in which, as predicted by Philip K. Dick [28], robots experience organicity, as delusion and/or certainty. The Actroid in *Comfortable and Alive* appears to know and feel affinity for "natural" and "organic" matter and phenomena.

> *See a profusion of red blooming flowers. See the bright red poppies growing near the ground.*
> *Smell the rich odor of red roses nearby. See red tulips opening to the morning sun. Experience*
> *the feeling of red all around. And as you walk forward through the deep red flowers, appreciate*
> *your own physical nature. Remember the physical sensations which bring you comfort and*
> *pleasure. Appreciate all of your physical senses, which allow you to be part of life and to expe-*
> *rience the fullness and joy of living.*
>
> Excerpt from *Comfortable and Alive*

The android dreams its connection to the storied human sensorium. If one takes seriously the idea that a cyborg bypasses the myth of genesis and is 'not subject to Foucault's biopolitics' [29, 65–67, 30, 190–195] – and I don't think I do, and I don't think Donna Haraway, author of the famed *Cyborg Manifesto*, does either – indeed, I believe her *Cyborg Manifesto* is an artwork – then it arrives paradoxically pure, exempt from cultural pressures, and ready to serve. But instead we must acknowledge that the cyborg is situated within humanity, sharing origins and connectivity [17, 31, 192–201]. Its cells, organic and computational, remember. (See studies in computational science, for example those cited in [32], on combining tissue engineering with android science.)

The major *diachronic* difference between biopolitical humans and cyborgs is argu-ably that a cyborg's factual and experiential memory is effectively acquired, and there-fore "located", in the present: a conduit for stored, pre-programmed, offsite information that flows directly through it, either modified or unmodified by algorithmic activity, when triggered by a human. In the whimsical milieu of *Comfortable and Alive*, the borg turns the tables, performing a hypnotic memory implant on its human audience. That is, it is stowing away experiential "memories", voiced in the language of the sensory encounter, inside another consciousness – also turning the tables, it might be able to trigger this human to retrieve these constructed and stored "memories" later, to whatever ends. In its professed empathic identification with – and perhaps longing for – 'the feeling of red all around' and 'the fullness and joy of living', the replicant is calculatedly twinning itself as memory tool with the consciousness it perceives but cannot remember.

6   Twinning and Replication

The onscreen hypnosis gynoid is presented in a twinned, *yin-yang* formation evoking (a) the above-described process of doubling and demonstration, (b) the overtly dialectic practice and discourse of geminoid replication, and (c) a schizophrenic postmodernity [33] or singularity wherein copy and original are indistinguishable one from another. The *yin* and *yang* in Confucian philosophy are dualistically figured, interdependent dynamics maintaining balance in the metaphysical and moral orders. However, in the Tao, borders between commonly assumed binaries represented by the *yin-yang* are perceptual, co-penetrative and never "real" [34]. Claudia Castañeda and Lucy Suchman [35, 9] write of the receding perceptibility of such borders due to the postmodern, post-human intensification of 'traffic between birthing and making'; this applies particularly to the humanoid robotics field wherein objects are made to be more and more like their models. Observing the experimental dualism in the *mise-en-scène* of *Comfortable and Alive*, it is tempting to ascribe difference to its two visualized versions of the Actroid in a kind of compare-and-contrast reflex. It's tempting, for example, to imagine one version as "master" persona and one as "slave". It is as tempting to personalize the entities as it is to anthropomorphize a machine; in this way we are conditioned to structure and make sense of each other and our creations [36, 36–37]. *Comfortable and Alive* provokes the conditioned dialectical aesthetic response in the gaze at the same time as it renders the response affective through the programmatic aspect of the text, which insists on the viewer's narrative subordination.

7   Conclusion

Rereading Freud through Derrida, Luce Irigaray in 'The blind spot of an old dream of symmetry' [37] addresses Freud's claim that little girls are malfunctioning little men.

> She argues that Freud could not understand women because he was influenced by the one-sex theory of his time (men exist and women are a variation of men), and expanded his own, male experience of the world into a general theory applicable to all humans. According to Irigaray, since Freud was unable to imagine another perspective, his reduction of women to male experience resulted in viewing women as defective men [38].

Of my works, *Comfortable and Alive* in particular seeks to interrogate this quasi-symmetrical composition and its implication of defective liveness and physicality, a "lack" residing in an existential netherworld. In this video piece, Actroid-F (or Gemi-noid-F) exists in this old, patriarchal blind spot in the posthuman collective dream (see also [17, 207]). It parodies, while contextualizing in contemporary robotics, what women have been culturally conditioned for centuries to do: to make others 'relax, and feel comfortable and alive'.

Acknowledgments. This work was supported by the Australian Federal Government through an Australian Postgraduate Award.

Comfortable and Alive has been shown in 2015 at UNSW Galleries, Sydney in the solo exhibition *Beyond Beyond the Valley of the Dolls*, and in 2016 at Gus Fisher Gallery, Auckland in the group exhibition *Alter*. For preview screeners, contact the artist at elenaknox.com. ·

References

1. Fukada, K.: Sayônara (Feature film). Phantom Film, Tokyo (2015)
2. Haring, K.S.: Gender differences paper. Personal e-mail, 12 June 2014
3. Nishio, S., Ishiguro, H., Hagita, N.: Geminoid: teleoperated android of an existing person. In: de Pina Filho, A.C. (ed.) Humanoid Robots, New Developments, pp. 343–352. I-Tech, Vienna (2007)
4. Ogawa, K., Taura, K., Ishiguro, H.: Possibilities of androids as poetry-reciting agent. In: Proceedings of 21st IEEE International Symposium on Robot and Human Interactive Communication, pp. 565–570 (2012)
5. Springer, C.: The pleasure of the interface. In: Wolmark, J. (ed.) Cybersexualities, pp. 34–54. Edinburgh University Press, Edinburgh (1999)
6. Matsumoto, Y.: Interview: incredibly realistic male and female android robots from Japan. In: Actroid, F.(ed.) DigInfo. https://www.youtube.com/watch?v=DF39Ygp53mQ. Accessed 2 Dec 2014
7. Guizzo, E.: Meet Geminoid F, a smiling female android. IEEE Spectrum, March 2010. http://spectrum.ieee.org/automaton/robotics/humanoids/040310-geminoid-f-hiroshi-ishiguro-unveils-new-smiling-female-android. Accessed 16 Mar 2016
8. Miraikan (National Museum of Emerging Science and Innovation). Android: What is Human? (Exhibition) (2014). http://www.miraikan.jst.go.jp/en/exhibition/future/robot/android.html. Accessed 16 Mar 2016
9. McCurry, J.: Erica, the 'most beautiful and intelligent' android, leads Japan's robot revolution. Guardian, December 2012. http://www.theguardian.com/technology/2015/dec/31/erica-the-most-beautiful-and-intelligent-android-ever-leads-japans-robot-revolution. Accessed 16 Mar 2016
10. Ahn, H.S., Lee, D-W., Choi, D., Lee, D-Y., Hur, M., Lee, H.: Uses of facial expressions of android head system according to gender and age. In: Proceedings of IEEE International Conference on Systems, Man, and Cybernetics, pp. 2300–2305 (2012)
11. T. Kim: Scientists test out all-robot dramas. Korea Times, February 2010. http://koreatimes.co.kr/www/news/tech/2010/02/129_60585.html. Accessed 16 Mar 2016
12. Wood, G.: Edison's Eve: A Magical History of the Quest for Mechanical Life. Alfred A. Knopf, New York (2002)
13. de Fren, A.: The exquisite corpse: disarticulations of the artificial female. Doctoral thesis, University of Southern California, Los Angeles (2008)
14. Frost, A.: Anatomical venus: the gory idealized beauty of wax medical models. http://dangerousminds.net/comments/anatomical_venus_the_gory_idealized_beauty_of_wax_medical_models. Accessed 16 Mar 2016
15. Kim. T.: Robot creators face design paradox. Korea Times, April 2007. http://koreatimes.co.kr/www/news/tech/2007/04/129_886.html. Accessed 16 Mar 2016
16. Carroll, C.: Us – and them. Natl. Geogr. Mag., August 2011. http://ngm.nationalgeographic.com/2011/08/robots/carroll-text. Accessed 16 Mar 2016
17. Suchman, L.A.: Human-Machine Reconfigurations: Plans and Situated Actions. Cambridge University Press, New York (2007)

18. Australian Broadcasting Commission: Japan unveils android newsreader. News, June 2014. http://www.abc.net.au/news/2014–06-25/an-japan-unveils-android-newsreader/5547886. Accessed 16 Mar 2016
19. Government of Japan (2007): Innovation 25: Creating the Future. Cabinet Office, Tokyo (2007). http://www.cao.go.jp/innovation/innovation/point.html. Accessed 26 Sep 2014
20. Robertson, J.: Gendering humanoid robots: robo-sexism in Japan. Body Soc. **16**(2), 1–36 (2010)
21. Butler, J.: Gender Trouble: Feminism and the Subversion of Identity. Routledge, New York (1990)
22. de Lauretis, T.: Technologies of Gender: Essays on Theory, Film, and Fiction. Indiana University Press, Bloomington (1987)
23. Halberstam, J.J.: Female Masculinity. Duke University Press, Durham (1998)
24. Aira, C.: Lovotics and the possibility of falling in love with robots someday. Silicone Angle, September 2013. http://siliconangle.com/blog/2013/09/20/lovotics-and-the-possibility-of-falling-in-love-with-robots-someday. Accessed 16 Mar 2016
25. Ueno, T.: Japanimation and techno-orientalism. In: Grenville, B. (ed.) The Uncanny: Experiments in Cyborg Culture, pp. 223–231. Vancouver Art Gallery & Arsenal Pulp Press, Vancouver (2001)
26. Levy, D.: Love and Sex with Robots. Harper Collins, New York (2007)
27. Oda, M.: Welcoming the libido of the technoids who haunt the junkyard of the Techno-Orient, or The uncanny experience of the post-techno-orientalist moment. In: Grenville, B. (ed.) The Uncanny: Experiments in Cyborg Culture, pp. 249–264. Vancouver Art Gallery & Arsenal Pulp Press, Vancouver (2001)
28. Dick, P.K.: Do Androids Dream of Electric Sheep?. Doubleday, New York (1968)
29. Haraway, D.: A manifesto for cyborgs: Science, technology, and socialist feminism in the 1980s. Socialist Rev. **80**, 65–107 (1985)
30. Cutler, R.L.: Warning: sheborgs/cyberfems rupture image-stream! In: Grenville, B. (ed.) The Uncanny: Experiments in Cyborg Culture, pp. 187–200. Vancouver Art Gallery & Arsenal Pulp Press, Vancouver (2001)
31. Haraway, D.: Simians, Cyborgs, and Women: The Reinvention of Nature. Free Association Books, London (1991)
32. Hasegawa, T., Collins, M.: The importance of random learning – Hiroshi Ishiguro interviewed by Cloud Lab. Volume Mag. **28**, 124–127 (2011)
33. Baudrillard, J.: The Anti-Aesthetic. Bay Press, Seattle (1991)
34. Ely, B.: Change and continuity: the influences of Taoist philosophy and cultural practices on contemporary art practice. Doctoral thesis, University of Western Sydney, Sydney (2009)
35. Castañeda, C., Suchman, L.A.: Robot visions. Soc. Stud. Sci. **44**(3), 315–341 (2014)
36. Ridgeway, C.L.: Framed by Gender: How Gender Inequality Persists in the Modern World. Oxford University Press, New York (2011)
37. Irigaray, L.: Speculum of the Other Woman. Cornell University Press, Ithaca (1974, 1985)
38. Donovan, S.K.: Luce Irigaray (1932–). Internet Encyclopedia of Philosophy (2005). http://www.iep.utm.edu/irigaray Accessed 16 Mar 2016

Robot Opera: A Gesamtkunstwerk for the 21st Century

Wade Marynowsky[✉], Julian Knowles[✉], and Andrew Frost[✉]

Macquarie University, Sydney, Australia
{wade.marynowsky,julian.knowles,andrew.frost}@mq.edu.au

Robot Opera proposes an avant-garde spectacle of performative media that places robots centre stage as signifiers of high culture within a 21st century total art work of the future. This chapter addresses how framing robotic performance as a *Gesamtkunstwerk* (and its historical ambitions) contributes to the canon of Cultural Robotics. The notion of robotic performance agency is detailed through the history and theories surrounding representations of the robot in popular culture, representations of robots as performance agents and through the dramaturgical concepts explored in Marynowsky's previous robotic art works.

Artistic [wo]man can only fully content [her] himself by uniting every branch of Art into the common Artwork: in every segregation of his [her] artistic faculties [s]he is unfree, not fully that which [s]he has power to be; whereas in the common Artwork [s]he is free, and fully that which [s]he has power to be [1].

The term 'Gesamtkunstwerk' was coined in 1927 by German philosopher Karl Friedrich Eusebius Trahndorff to describe the concept of the 'total art work' - a work which synthesises all art forms into a single unified multidisciplinary work [2]. The term is most closely associated with the works of Richard Wagner (1813–1883), who sought to draw on the concept to guide his experiments in the opera form. In his text, 'The Artwork of the Future' (1849), Wagner calls for a synthesis of all artforms to produce the total art work of the future by pursuing musical drama (opera) as an integrating structure [1]. The concept of the total art work, drawing on multiple disciplinary practices, histories and conceptual reference points, is one which aligns very closely with the history of media art. One could say the Gesamtkunstwerk has found its natural home within the realms of contemporary media art. As a consequence, one can draw direct parallels between the Wagnerian approach to opera and the emerging conditions of mediatized and robotic performance agency and pose the question as to whether robotic opera may be seen as the logical playing out of the historical ambitions for the Gesamtkunstwerk within the opera tradition.

This chapter explores this proposition through the work Robot Opera (2015), a robotic opera for eight semi-autonomous robot performers. The work has been realised by Wade Marynowsky (robotic artist) in collaboration with Julian Knowles (music/ sound) and Branch Nebula, Mirabelle Wouters and Lee Wilson (lighting, dramaturgy). Informed by the underlying fields of creative robotics, mediatized performance, music, and interactive media art, the project merges artist driven algorithmic/choreographic concepts with audience driven agency within a large scale performance interaction space 42 × 25 m. The project brings together core areas of investigation within these

© Springer International Publishing Switzerland 2016
J.T.K.V. Koh et al. (Eds.): Cultural Robotics 2015, LNAI 9549, pp. 143–158, 2016.
DOI: 10.1007/978-3-319-42945-8_12

disciplines by establishing a performative context to explore the concept of robotic performance agency.

In order to understand how robots might operate as performers within an operatic performance context it is necessary to understand the histories and theories of robots through their representations in popular culture as well as their representations as performance agents.

1 Representations of Robots in Popular Culture

The origins of the robot in Western popular culture can be traced to the early 19th century. Stableford and Langford cite early clockwork dummies and other mechanised puppets as key influences for the mechanical beings that appear in E.T.A Hoffmann's stories *Automata* (1814) [3] and *The Sandman* (1817) [4], characters such as the 'Talking Turk' and 'Olympia' that "present a [...] verisimilitudinous image, and play a sinister role, their wondrous artifice being seen as something blasphemous and diabolically inspired" [5]. The inherently ambiguous nature of the literary and, later, the cinematic robot has proven to be as durable as the figure of the robot itself, a mixture of techno-logical wonder and uncanny dread, an often ill-defined amalgam of the mechanical being (the robot), artificial intelligence (the computer), the human-machine hybrid (the cyborg) or human simulacra (the android).

The nature of these robots and robot-like beings depend on the requirements of the stories in which they appear. Robots such as 'Robby' in Fred M. Wilcox's *Forbidden Planet* (1956) [6] or his Soviet counterpart 'John' in Pavel Klushantsev's *Planeta Bur* (1962) [7] dutifully follow Isaac Asimov's *Three Laws of Robotics* (1941) [8] sacri-ficing themselves to save humans. This kind of neutral 'goodness' is contrasted by robots such as the android Gunslinger in Michael Crichton's *Westworld* (1973) [9] or the genocidal Cylons in *Battlestar Galactica* [10], just two examples of a widely held conception of the robot as inhuman machine where the first law of robotics is blatantly broken: "a robot may not injure a human being or, through inaction, allow a human being to come to harm" [11].

So where does this ambivalent-neither friend nor foe-view of the robot derive? Robert M. Geraci traces the historical origins of robots beyond the early decades of the 19th century to the ancient Greek Myths of "Pygmalion and Daedalus to the Jewish Golem and the homunculi of Renaissance alchemy" [12] For Geraci, "The Western goal of building a functional humanoid also received, no doubt, some of its impetus from religion" [12]. From homunculi to singularity theories (Neuman, Kurzweil) the creation of robots and artificial intelligence (AI) may be considered an act of the divine, but at the same time, a mortal sin from a theological perspective. Once the creation of a humanoid by humans is achieved, then the end of the world is nigh. Western audi-ences have been easily swayed by the fears of technology found in Karel Capek's *R.U.R, Rossum's Universal Robots* (1921) the origin of the modern robot story and the source of the word 'robot' itself "...derived from the Czech *robota* (statute labour)" [5]. As performance theorist Steve Dixon states, Capek's play "concerns the supplanting of humans by robots and has been discussed as a warning against Frank-ensteinian scientific hubris" [13].

The legacy of the 'Golden Age' of science fiction magazine publishing - roughly from the early 1930s to the mid 1960s - and its overlap into cinematic science fiction from the early 1950s onwards - produced a vast cultural trove of images of the robot that have proven remarkably durable. The 'mechanical man' image of the robot was in part established by artists such as Frank R. Paul, Robert Fuqua, Ed Emshwiller and Virgil Finlay producing illustrations for magazines such as *Astounding*, *Amazing* and the *Magazine of Fantasy and Science Fiction* that provided the basis for the design of cine-matic robots such as the aforementioned 'Robby' and 'John' or the silver and sleek 'Gort' seen in Robert Wise's *The Day The Earth Stood Still* (1951) [14]. The non-humanoid robot is, by comparison, a rare sight in cinematic visions of mechanical intel-ligence, either human-made or alien: the alien machine of *Kronos* [15] is a gigantic black cube with cylindrical legs that rampages around the Earth in search of energy, a strange anticipation of minimalist sculpture of the 1960s and the alien artifact in Stanley Kubrick's *2001: A Space Odyssey* (1968) [16], yet it is an outlier in conceptions of the 'look' of the robot in science fiction cinema.

More recent science fiction films such as *I, Robot* [17] - an adaptation of Asimov's *Robot* stories - present the robot as artificially intelligent machine with a speculative design based on the first generation frosted-plastic iMac, and an equally familiar homi-cidal mission thanks to some sinister covert reprograming. If Proyas's film posits the robot as a logical extension of contemporary consumer electronics then Alex Garland's *Ex Machina* (2015) [18] is an example of the popular conception of a robot that is indistinguishable from a human, if only when judged by outward appearances. Attempts to create robots in real life have often met with the same problems that filmmakers have encountered attempting to exactly simulate humans onscreen by means of computer animation - the 'uncanny valley' [19] (discussed in The Uncanny below).

Representations of Robots as Performance Agents. Jean Tinguely's 'Painting machines or Metamatic sculptures' (1959) are autonomous machines that paint pictures. The agency displayed in these works parodies the human thought processes needed to produce an abstract expressionist painting. Tinguely's work suggests that once the orig-inal concept is conceived by a human, then a machine can take over in the process of fabrication - but at what stage can a machine be perceived to produce original thought? Or at least to be able to *perform* convincing agency? This section explores the notion of robotic performance agency through the disciplines of the visual arts, music and theatre. The robotic agency, aesthetic principles and the context in which the robot is presented can help us understand the liminality of the performative robot, where and when it becomes an acceptable representation and or generator of 'living' culture.

The cybernetic sculptures of Edward Ihnatowicz such as his *Sound Activated Mobile* (SAM) (1968) and *The Senster* (1970) can be understood as distant robot relatives, precursors to contemporary robotic artworks, for example the works of Bill Vorn and Louis Phillipe Demers. *The Senster* was a large, steel, two legged zoomorphic creature that had a moving arm with multiple degrees of freedom. The arm's movements reacted to people's voices (via microphones) and to their movements (via radar), "the rest of the structure would follow them in stages if the sound persisted. Sudden movements or loud noises would make it shy away" [20]. One of the first kinetic sculptures to be computer controlled *The Senster* was commissioned by electronics company Phillips and exhibited in The Evoluon, a remarkable flying saucer shaped building in Eindhoven, The Netherlands. *The Senster* has informed the main directions of robotic art, through the way it responded to its audience, with its animal-like behaviour and machine aesthetic.

Ihnatowicz's legacy and machine aesthetic can be seen in the works of Vorn and Demers, for example Vorn's *Hysterical machines* (2006) that can be read as zoomorphic mechanical spiders that hang from the ceiling. The machines have a spherical body and eight moving arms made from aluminium tubing and electronics. They have a "sensing system, a motor control system and a control system that functions as an autonomous nervous system (entirely reactive)... the perceived emergent behaviours of these machines engender a multiplicity of interpretations based on a single dynamic pattern of events" [21]. The robotic performance agency in both of these works (the natural fluidity of *The Senster's* arm movement and twitching arms in Vorn's *Hysterical machines*) generates a similar response, a temporary zone for reciprocity between the artificial and the human, known as the field of Human-Robot Interaction (HRI). Put simply, robots may react to human presence and humans project their internal desires onto the simulacra. Through the use of stark lighting and an eerie soundtrack, Vorn dramaturgically sets the scene for the audience, in order to highlight our desire to anthropomorphize his articulated metal structures, his aim being to induce empathy for his robotic creations.

Similarly, prominent electronica musician Tom Jenkinson (aka Squarepusher) felt empathy for the musical androids he collaborated with, Z-machines: "the robots [are] sad because they are just treated by the public as entertainment machines... their other qualities are neglected...this sadness comes out in the music they play...strangely [this] becomes one of the reasons why the public likes them, because they seem to be able to evoke strong emotions in their audience" [22] Z-machines consists of "March, a 78-fingered guitarist; Ashura, a drummer with 22 arms; and Cosmo, a keyboardist who triggers notes with lasers" [22]. The performance agency of the Z-machines can be understood as extending music beyond that which is physically possible for human players. By creating super-human compositions that are played faultlessly and easily reproduced evokes strong emotions in humans, as we feel threatened by being replaced by machines. In this example, it may be understood that the Z-machines androids are creators of *culture* as they play as humans do in the social formation of a musical ensemble.

In contrast to the slick techno-fetishtic finish of the Japanese Z-machines is the steampunk aesthetic of the Berlin based Compressorhead. The android music ensemble features three band members built to human scale: 'Fingers', the guitarist; 'Stickboy', the drummer; 'Bones', the bassist and 'Junior', the hi-hat humper [23]. The group perform cover versions of well known repertoire from the heavy metal canon, such as Motorhead's *The Ace of Spades* and Joan Jetts' *I love rock and roll*. Compressorhead have been touring Europe and Australia since 2012, performing their one hour gig to large crowds normally expected at rock concerts, for example, the Big Day Out, Sydney, 2013. The robotic performance agency experienced when being entertained by

Compressorhead is convincing because they can actually play the well-known songs they are programmed to play, at the same time they make dance-like gestures, head-banging and swaying side to side. Compressorhead are successful representation of cultural robots existing in the rock and roll context they were created for. At this stage Z-machines and Compressorhead are simply midi control devices that actuate pre-written musical scores. Until the bands write their own material through machine learning algorithms they are not considered to be 'creators' of culture.

Using new media dramaturgical concepts in combination with the traditions of the stage roboticist Hiroshi Ishiguro and playwright/director Oriza Hirata have created several theatre works using Ishiguro's robots. I, Worker (2008) with Wakamaru, a humanoid robot, *In the Heart of a Forest* (2010), featuring Wakamaru), *Goodbye* (2010), with Geminoid F, a female android, and *Three Sisters*, *Android Version* (2012), with Wakamaru and Geminoid F. Wakamaru is programmed to move and talk when performing its role, while its operator controls the timing of the robot's actions remotely. Geminoid F is also controlled tele-remotely by a female actor who operates it. The robotic performance agency experienced through the robotic characters invites an empathetic response, similar to that of real actors on stage.

2   Dramaturgical Concepts in Marynowsky's Previous Robotic Work

The difference in the above examples of robots as performance agents and Marynowsky's investigations is that, in Marynowsky's work the audience is invited to directly engage and interact with the robots, within a gallery space. This breakdown of the fourth wall (an invisible barrier between the performer and the audience) is a key concept in the western avant-garde traditions of performance art. For example, Alan Kaprow's 'Happenings' in which, the audience participation in the performance directly affected its outcome. Thus, Marynowsky's works draws connections between nineteen sixty's conceptual and performance art and art in the age of robotic performance agency.

The scale and the agency in Marynowsky's robotic work can often be threatening, with large robots travelling towards the audience, they must make their own decisions as to either move out of the way or hope the robot stops before colliding with them. This intimidating experience draws on the work of La Fura Das Bas, a Spanish performance art group who took the notion of 'Happenings' and performance art to the next level, by controlling their audience in often threatening ways. In doing so, they blurred the line between the performer and the audience. As academic Maria Delgado states, "One does not watch a performance of La Fura. One participates" [24].

Robot Opera also extends upon the dramaturgical concepts explored by Marynowsky over the past two decades emerging from the context of the visual arts. These dramaturgical concepts include: The Uncanny; The Camp; The Robot as High Culture and are framed within different models of audience Reciprocity.

3   The Uncanny

The uncanny is a key concept in western humanities as well as in android science. Psychoanalyst Ernst Jentsch states that a very good instance of the uncanny casts "doubts [as to] whether an apparently animate being is really alive or conversely, whether a lifeless object might not be in fact animate" [25]. He lists waxwork figures, ingeniously constructed dolls and automata to have the potential to invoke an uncanny impression. Further to this, cultural theorist Terry Castle argues that the eighteenth century invention of the automaton was also the invention of the uncanny [26]. Sigmund Freud sought to further Jentsch's definition, proposing that 'the uncanny' is "what is frightening – what arouses dread and horror; equally too, the word is not always used in a clearly definable sense" [27]. Freud thus proposes that the uncanny has a role in eliciting emotional reactions from humans. The uncanny can be understood as an eerie, mysterious and weird feeling that extends beyond what is normal or expected, often-suggesting superhuman or supernatural powers or qualities.

The uncanny continues to be an enduring concept in visual arts. In the 1920's, the surrealists' love of the automaton was subconsciously explored through repressed desire. Hal Foster [28] understands Freud's investigation of the uncanny as the core conceptual undercurrent in the Surrealist movement in his book *Compulsive Beauty*. While Bruce Grenville relates the uncanny to notions of the cyborg in popular culture and aesthetics, he argues that Marcel Duchamp's *Nude Descending a Staircase* (1912) is a representation of the uncanny human-machine in motion, considered at its time "not only a threat to popular aesthetics but also a threat to the popular public perception of the human body and its physical limits" [29].

The artist Mike Kelley was "struck by Jentsch's list and how much it corresponded to a recent sculptural trend - popularly referred to in art circles as mannequin art" [30], and began collecting images of this type of work, later forming the major exhibition, *The Uncanny*, at Tate Liverpool in 2004. The exhibition consisted of life-sized figurative sculptures from throughout the ages, all with a disturbing edge: Hans Bellmer's *Doll* (1936), the Andy Warhol robot (1981), Disney's animated audio-animatronic figure of Abraham Lincoln (1964), mannequin stand-ins for the influential electronic band Kraftwerk (1978), as well as medical models and images of Jacques de Vaucanson's automata, such as his defecating robot duck (1739). For many artists, the uncanny continues to be a desired effect, for example, the works of Ron Mucek, Damien Hirst, Paul McCarthy, Tony Oursler and Patricia Piccinini, to name a few. Whilst the uncanny can be found across a range of visual art forms it finds its most potent expression in the field of robotics.

Robotics scientist Masahiro Mori proposed the 'uncanny valley' hypothesis [19] as the relationship between human likeness and perceived familiarity: "familiarity increases with human likeness until a point is reached at which subtle differences in appearance and behaviour create an unnerving effect" [31]. Following in Mori's footsteps, Karl. F MacDorman theorised that the android in the 'uncanny valley' elicits an eerie sensation because it is acting as a "reminder of mortality" [31]. For Mori, movement amplifies this effect and he "cautioned robot designers not to make the second peak their goal – that is, total human likeness – but rather the first peak of humanoid appearance to avoid the risk of falling into the valley" [31]. If we accept Mori's hypothesis the 'uncanny valley' can only ever be overcome when a truly humanoid robot (indistinguishable from a human) is produced, until which time we can only speculate through both an artistically and scientifically driven liminality.

In Marynowsky's prior works, the uncanny is embraced as an overall aesthetic - a device to invite the viewers into conceiving of the robots as beings that exist in their own right. Once the unnerving part of uncanny experience is overcome the human-robot experience can be opened up to various other more rewarding interpretations and experiences.

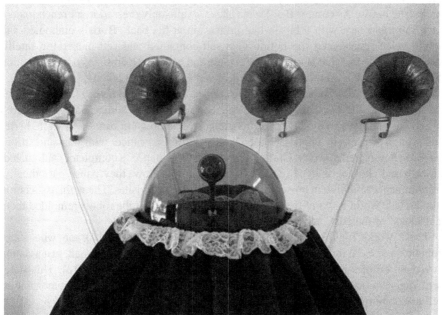

Wade Marynowsky, *The Discreet Charm Of The Bourgeoisie Robot,* 2008.

4   The Camp

A number of Marynowsky's prior works have investigated notions of 'camp' in respect of robotic identity. The robots are charged with an affectation that challenges gender based stereotypes of his android characters. The notion of camp is explored via the use of robots 'in drag', using the symbols of transvestism to confuse gender roles assigned by humans to robots. Susan Sontag states that camp is "the consistently aesthetic experience of the world. It incarnates a victory of 'style' over 'content', 'aesthetics' over 'morality', of irony over tragedy" [32] and its key proponents are "an improvised self-elected class, mainly homosexuals, who constitute themselves as aristocrats of taste" [32]. Sontag identifies the theatricalisation of experience, the exaggeration of mannerisms and the deployment of irony as key components of camp. The notion of camp is this strongly tied to human experience and behavioural codes. As such, camp is a powerful device to inflect the robot with human qualities.

In *The Discreet Charm Of The Bourgeoisie Robot* (2008) for the length of the three week exhibition, the artist acted in a carefully choreographed drama as *Boris, The Bourgeoisie Robot*. A bricolage of 1950's science fiction (web camera under a domed head) and Victorian dress (body), the robot avatar waited for an audience to enter the gallery and then conversed with them in a polite and pleasant manner. The voice (filtered through a vocoder) was pure computer coldness inflected with the accent of an upper class English toff. The conversations covered champagne, caviar and were quickly re-directed to topics about itself, setting the scene for an interaction with a narcissistic entity. As curator Bec Dean states, "exquisitely dressed in a French maid's black satin and lace with a bustle-like protrusion at his back, Boris's embodied and mobile voice represented the notion of a self-contained and self-preserving intelligence" [33]. The fact that Boris's voice is male and he wears a dress introduces a notion of camp, as does his adoption of a theatricalised English toff persona.

The Hosts: A Masquerade of Improvising Automatons (The Hosts) (2009) features five larger than life sized robot characters. The robots wear sumptuous embroidered ball gowns and have individual masquerade guises: a clown in black and white harlequin print; a princess in a pink-ribboned bodice; a military officer with stars and stripes and a cowboy-hatted cowboy. Gliding gracefully, they 'dance' a completely automated, sensor-based choreography. Lights dimming to a dull glow, they pause periodically, and commence spinning in unison like robotic whirling dervishes. The main association people made while walking freely among the work, was that the robots reminded them of 'Daleks' of the BBC series *Dr.Who*, or 'Daleks in Drag' [34].

These works highlight the camp robot concept theorised by Dixon who states "robotic movement mimics and exaggerates but never achieves the human, just as camp movement mimics and exaggerates but never achieves womanhood" [13]. This statement suggests that camp is an essential factor in understanding anthropomorphic as well as zoomorphic robot performance agency.

Wade Marynowsky, *The Hosts: A Masquerade of Improvising Automatons*, 2009.

5   The Robot as High Culture

The notion of the camp robot is also linked to the idea of high culture but with an ironic twist "the experiences of camp are based on the great discovery that the sensibility of high culture has no monopoly upon refinement…the whole point of camp is to dethrone the serious, camp is playful" [32]. The camp not only enables imparting human qualities to robots, but it assists them to become credible agents within a high culture context. Once placed in the context of the art gallery robots immediately become accepted as fine art. This, concept was introduced by Marcel Duchamp when he placed a Urinal in the gallery and signed it R. Mutt (1917) or when Andy Warhol painted his Campbell's soup cans (1962). Importantly the avant-garde tradition of the visual arts involves re-writing what has gone before, re-defining what art is. This dissonance attempts to break down elite systems of class and hierarchy, namely that of bourgeoisie society. This is exemplified in Marynowsky's work *The Discreet Charm Of The Bourgeoisie Robot* (2008), which ironically suggests that high art is for robots.

The literary fiction devised by ETA Hoffman in the early 19th century popularised the darker side of the automaton. However most people never actually physically experienced an autonomous performance, as automata existed mainly for the courts of royal society and eventually those who could spare a week's wage. 18th Century automata were hand-built custom-made marvels created by clock-makers and mechanics, with Jacquet Droz of Switzerland and Jacques de Vaucanson of France being two of most-cited automata makers of the time. Vaucanson "achieved most notoriety as a producer of a high-society spectacle…[with] his magnificent creations…praised by kings and applauded by scientists" [35]. Presenting to the Académie of Sciences in Paris, 1738, Vaucanson "set the standard for mechanical androids" [36] with his flute player, a drum player and a digesting duck. The life sized, life-like and musical qualities of his automatons gained "the attention of influential people such as Voltaire, Frederick the Great and the general minister to Louis XV" [36]. Fortunately the works of Droz and Vaucanson are still functioning and are regularly on display in Musée d'art et d'histoire in Neuctel, Switzerland and the Conservatorie national des arts et métiers in Paris, preserved through the conservation of culture, the automatons remain representations of the periods intelligentsia.

An important point to this argument (to accept the robot or more definitively, the automaton as a signifier high culture), is detailed in the book *Androids in the Enlightenment: Mechanics, Artisans, and Cultures of the Self* in which, Associate Professor Adelheid Voskuhl proposes, that automata were harbingers of the burgeoning industrial age, an age where the automaton transformed into the popular notion and understanding of the robot. A robot in this context is a product of mass-production, dispersed to the masses as popular culture, whereas automatons are one-of-a-kind artist's creations, preserving the aura of the art object.

In the author's experience, the automaton has re-emerged as a credible participant in high culture, expanding the status of Cultural Robotics through representation in major international contemporary art biennales (biennales being the cultural signifiers of international contemporaneity for the hosting nation). For example Marynowsky's work *The Hosts* featured in Beyond Mediations, Mediations Bienniale, The International Biennale of Contemporary Art, Poland, in 2010. The work's placement into the circular ballroom in the clock tower of the Imperial castle (Zamek) strategically situated the work in the traditions of European antiquity. This suited the project aptly as the work drew its inspiration from the traditions of 18th Century European automata and their fashion. In 2014, Marynowsky's *The Acconci Robot* featured in thingWorld: International Triennial of New Media Art, National Art Museum of China. By hosting the exhibition Chinese nationals are declaring how forward thinking and innovative they are by accepting new media art into their National Art Museum.

The tradition of displaying robots as representations of high or intelligent culture to display technological ingenuity is foregrounded in the meeting of the President of the United States Barack Obama and Honda's Asimo in Japan, 2013. After initial greetings and a display of dexterity (with Asimo jumping up and down on one leg), Obama and Asimo kicked a soccer ball back and forth. The experience left Obama with an uncanny feeling, stating, "I have to say that the robots were a little scary, they were too lifelike, they were amazing" he said [37].

6 Reciprocity

If the goal of android science is to advance human-robot relationships and to find adequate design concepts to support meaningful interactions, then artists are able to subvert, pervert and critique these notions through experimentation, within the direct, open ended context of the art gallery. A range of Marynowsky's prior works have explored different models for Human Robot Interaction. Notions of *reciprocal* exchange are explored in *The Discreet Charm Of The Bourgeoisie Robot* (2008), and *non-reciprocal* exchange in *The Hosts* (2009). Whilst *The Acconci Robot* (2012), subverts audience expectations of the direct engagement and reciprocal exchange by creating a robotic character that only responds when the audience is most *disengaged* from it.

Audience responses have provided insights into the way in which robotic agency was read in the context of an artwork. *The Hosts* demonstrated that audiences experiencing robots in the context of the gallery are desperate for reciprocal exchange. They seek feedback from robots in human-like ways, as might be expected from literary or cinematic science fiction genres. They wave their hands at the robots 'heads' and mimic the robots' movement. The main psychological response to *The Hosts* was that the robots responded to humans personally, when in fact they are autonomous. They simply avoid any obstacles in their path whilst wandering, with obstacles triggering pre-programmed sound and light samples.

As a viewer of the work, Melody Willis recalled, "They all turned and gathered around me. I felt psychically powerful, like a child with extrasensory perception (ESP), but then they started spinning madly and I realised I was meddling with forces I could never understand" [38]. In Willis's account, she expresses that she "could never understand" because the robots she thinks she is controlling with her mind, are in fact acting autonomously, ignoring her. The lack of reciprocal exchange between robots and humans causes uncertainty as to what the robot is thinking. As social robotics researcher Pericle Salvini explains: "the lack of presence causes uncertainty, especially when a physical entity gives the impression that there is more behind it, that there is indeed something behind the mask" [39].

In *The Discreet Charm Of The Bourgeoisie Robot* Marynowsky attempted to convince viewers of the robot's intelligence through tele-operation, by remotely controlling the movement and voice of the robot. The elaborately costumed robot avatar waits for an audience member to enter the gallery and converses with them in a camp and narcissistic manner adopting a model of direct reciprocity with the audience. The reciprocal exchange caused an ambiguous response amongst audience members. There was no uncertainty as to whether the robot was alive or not, but how intelligent was it and how could it be so intelligent? The Human Robot Interaction in the work became a game of interrogation between the robot and humans, a drag form of the Turing Test [40]. Dan McKinlay states in his review, "the conversation invokes and inverts that old new-media parody, the *ELIZA* [41] psychoanalysis program" [42].

Wade Marynowsky, *The Acconci Robot*, 2012, photo by Mark Ashkanasy, RMIT Gallery.

Marynowsky's work *The Acconci Robot* is an interactive robotic character that follows the viewer when they are not looking at it. Appearing as a shipping crate of minimal design, the robot is mute and motionless as a viewer approaches. But when the audience member turns away, and starts to leave, the robot begins to follow. If the audience member turns to look back at the robot, it stops in its tracks. The work draws inspiration from the 1969 performance work, 'Following piece' by Vito Acconci [43]. Acconci's early work was developed from an interest in the human body and its relationship to public space. In 'Following Piece', Acconci would select unsuspecting people in the street and follow them until they disappeared into a private place. Acconci carried out this performance every day for a month, documenting each encounter and sending it to a different member of the arts community. Acconci's investigations of the body in public space are re-contextualised in the work within the gallery context, re-examining public interaction through Human Robot Interaction. The work explores the concept of an anti-reciprocity through recognition of the human as the subject of surveillance, through the act of following. Leading the viewer to question notions of robotic agency, an important aspect of the increasingly computer mediated times we live in, for example Unmanned Aerial Vehicles (UAVs) and drones.

7   Robot Opera and the Exploration of a Robotic Performance Practice

Building on the dramaturgical concepts in Marynowsky's previous works: The Uncanny; The Camp; The Robot as High Culture and audience Reciprocity, *Robot Opera* moves beyond gallery based installation contexts and deploys robots as perform-ance agents in an operatic context. In Western culture, opera is seen as a strong symbol of class and is framed as the pinnacle of high culture and is a heavily stylised performance form with a range of identifiable performance conventions. This makes opera a fertile site to investigate the potential for robots to be seen as performance agents and whether there is the potential to conceptualise a 'performance practice' that extends beyond their more traditional role as automated devices executing recorded sequences.

It is notable that opera as a form has remained fairly stable and somewhat resistant to radical transformation. As Salter asserts "Despite the interest in expanding the musical language of opera through the new compositional languages arising from serialism and postmodernism [including minimalism], many of these attempts still retained the dramatic stagecraft and orchestral vocabularies of traditional opera" [44]. Opera has therefore not tended to be a site for radical transformation. Furthermore more, radical engagements with the form of opera form have tended to come from outside the classical music field. Nam June Paik sign posted opera in his work *Robot Opera* from 1964. Paik had developed a robot, named K-456 (named after a Mozart piano concerto), in the early 1960s that become the focus for a range of subsequent art works and happenings. K-456 was anthropomorphic in appearance, was radio controlled, played audiotaped speeches by John F. Kennedy and defecated beans. K-264 had its first public performance in 1964 in Paik's own *Robot Opera* with Paik and Charlotte Moorman. Despite what the title of the work may suggest, Paik's piece had more to do with the avant garde Happenings of the period and did not expressly reference or draw upon opera as a form.

Perhaps the most important precedent work to *Robot Opera* is Tod Machover's *Death and the Powers* (2010) [45] developed via the MIT Media Lab. This large scale work involves computer controlled set elements and autonomous robots alongside human performers. Machover achieves a very high degree of sophistication in respect of the dram-aturgical treatment of robotic performers. Furthermore unlike Paik, Machover's work directly addresses the opera tradition and has proven to be perceived as a work within that canon. A range of performances have been staged by large mainstream opera companies and the work can be seen to have entered the operatic repertoire. The key difference in respect of the *Robot Opera* project is that *Death and the Powers* relies on human performers as singers/actors and the robotic elements are supplementary to a human cohort of performers. The work does not solely rely on the performative agency of the robot performers or rest upon an entirely robotic dramaturgical setting. The work therefore provides insights into mixed cohort (robot/human) performance and the realm of the robot-only operatic performance remains unexplored. *Robot Opera* seeks to explore this mode of performance and develop a notion of robotic performance practice.

Robot Opera features eight larger than life sized rectangle monolithic shaped robots on powered wheels, employing the machine aesthetic (of Ihnatowicz's The Senster) the robots are equipped with Kinect v2 cameras that allow the robots to respond to humans by translating their proximity and facial expression into responsively programmed sound and light, on the robots. The robots are individual agents operating on a wireless network and operate according to algorithmic principles, with various choreographed behaviours executed from the robot performance cohort, alongside sensing systems that allow the robot to be responsive to audience behaviours and interventions. The robots are not explicitly humanoid in appearance, but incorporate anthropomorphic design principles - for example, Kinect cameras for eyes, loudspeakers for mouths, and sensor systems to detect others. The work thus achieves a sense of the uncanny and the ambiguity of liveness without resorting to explicit humanoid representation.

Unlike much installation work, performance work most often deploys specific time-based structuring principles. It is 'vectorised' in the temporal domain, in that, performance works are perceived as having a beginning, middle, and an end. Notions of development exist and there is often a dramaturgical shape, or at the least, a sense of a set schema in respect of the performance structure and content. *Robot Opera* seeks to explore the idea of the robot as an active agent with the performance context, moving beyond a programmed machine executing digital sequences towards a semi-autonomous state, where the robots are seen to execute context specific decisions based on Human-Robot Interaction. This robotic performance agency can be distinguished from the fields of interactive or algorithmic art more broadly, in that it is explicitly situated within a performance context and so invites the audience to consider the robots as performance agents within a performative and dramaturgical system making 'performance decisions'.

Wade Marynowsky, *Robot Opera,* 2015, Photo: Heidrun Lohr, Carriageworks Sydney.

Within *Robot Opera* such a schema exists in the form of software based control sequences and behaviour commands that are plotted against a timeline. The performance model therefore incorporates the idea of a script of choreography, but allows for the audience responses and features of the performance space to modulate and inflect the pre-determined script/choreography. In so doing the robotic cohort starts to model a human cohort working to a script or choreography but having the freedom to inflect the performances based on audience and site conditions. The sense of the anthropomorphic extends beyond the physical attributes of the robotic form to the behaviour in performance. On a conceptual level, this constructs a set of relations in robotic performance that map onto a human performance paradigm. In human performance, the schema is mediated to varying degrees by the performance context - that is, the limitations and possibilities of the venue/site and the audience real-time responses, be those subtle or unsubtle. These elements have a structuring effect on performance and this connects deeply to the fundamental concept of what a performance is.

By modeling the robotic performance system on human performance paradigm then the robots can be experienced as performers in their own right and not be seen to be sequencers or machines, executing patterns that pay little regard to their context. The project therefore suggests that the notion of robotic performance agency can be identified from the arising technical approach and the performance context. It is proposed that this form of agency is specific to performance-based robotics because it invites the audience to consider the robots as performance agents within a performative and dramaturgical system making 'performance decisions'. *Robot Opera,* then, is a work that opens up the possibility of a new robotic performance practice, expanding the field of Cultural Robotics. Placing robots centre stage as signifiers of high culture within a 21st century total art work of the future.

References

1. Wagner, R.: The Art-Work of the Future. Trans. William Ashton Ellis. The Wagner Library
2. Trandorff, K.: Aesthetik oder Lehre von der Weltanschauung und Kunst. Berlin, Maurer (1827)
3. Hoffmann, E.T.A.: Automata (2006). ReadHowYouWant.com
4. Hoffmann, E.T.A.: The Golden Pot and Other Tales. Oxford University Press, New York (2000)
5. Stableford, B., Langford, D.: 'Robots', Encyclopedia of Science Fiction (2015). http://www.sf-encyclopedia.com/entry/robots. Accessed 09 June 2015
6. Forbidden Planet. [film] Hollywood: Fred M. Wilcox (1956)
7. Planeta Bur. [film] Leningrad: Pavel Klushantsev (1962)
8. Asimov, I.: Reason. Astounding, 33–55 (1941)
9. Westworld. [film] Hollywood: Michael Crichton (1973)
10. Battlestar Galactica. [TV program] Sci-Fi Channel: NBC Universal Television (2004–2009)
11. Clarke, R.: Asimov's Laws of Robotics: Implications for Information Technology. Cambridge University Press, Cambridge (1993)
12. Geraci, R.: Apocalyptic AI: Visions of Heaven in Robotics, Artificial Intelligence, and Virtual Reality. Oxford University Press, New York (2010)
13. Dixon, S.: Metal performance humanizing robots, returning to nature, and camping about. TDR Drama Rev. **48**(4), 15–46 (2004)
14. The Day The Earth Stood Still. [film] Hollywood: Robert Wise (1951)
15. Kronos. [film] Hollywood: Kim Neumann (1957)
16. A Space Odyssey (2001). [film] Hollywood: Stanley Kubrick (1968)
17. I, Robot. [DVD] Hollywood: Alex Proyas (2004)
18. Ex Machina. [film] London UK: Alex Garland (2015)
19. Mori, M.: The uncanny valley. Energy **7**(4), 33–35 (1970). http://spectrum.ieee.org/automaton/robotics/humanoids/the-uncanny-valley. Accessed 14 June 2015
20. Zivanovic, A.: http://www.senster.com/ihnatowicz/senster/. Accessed 14 June 2015

21. Vorn, B.: http://www.billvorn.ca. Accessed 14 June 2015

22. Bakare, L.: http://www.theguardian.com/music/2014/apr/04/squarepusher-z-machines-music-for-robots. Accessed 27 July 2015

23. Kolb, M.: http://compressorhead.rocks. Accessed 27 July 2015

24. Antonio, F., de Queiroz, P.V.: Will all the world be a stage? The Big Opera Mundi of La Fura dels Baus. Romance Q. **46**(4), 248–254 (1999)

25. Jentsch, E.: On the psychology of the uncanny. Angelaki, Urbana-Champaign, trans. Roy Sellars, p. 11 (1906)

26. Castle, T.: The female thermometer: eighteenth-century culture and the invention of the uncanny, p. 11. Oxford University Press, New York (1995)

27. Freud, S., Strachey, J.: Art and literature: Jensen's Gravida, Leonardo da Vinci and others works. Penguin, Harmondsworth (1985)

28. Foster, H.: Compulsive Beauty. MIT Press, Cambridge (1995)

29. Grenville, B.: The Uncanny: Experiments in Cyborg Culture. Arsenal Pulp Press, Vancouver (2001). p. 10, 18, 37

30. Kelley, M., Welchman, J.C., Grunenberg, C.: The Uncanny (2004): by Mike Kelley, artist; [on the occasion of the Exhibition The Uncanny by Mike Kelley, Tate Liverpool, 20 February – 3 May 2004; Museum Moderner Kunst Stiftung Ludwig Wien, 15 July – 31 October 2004]

31. MacDorman, K.F.: Androids as an Experimental Apparatus: Why is there a unanny valley and can we exploit it? In: Android Science, a Cog Sci 2005 Workshop, Osaka, Japan (2005). http://www.androidscience.com/proceedings2005/MacDormanCogSci2005AS.pdf. Accessed 14 June 2015

32. Sontag, S.: Notes on camp. In: Camp: Queer aesthetics and the performing subject, pp. 53–65 (1964)

33. Dean, B.: The Hosts: A Masquerade of Improvising Automatons. Performance Space, Sydney (2009)

34. The term Dalek in Drag was communicated by the audience to the artist Wade Marynowsky in relation to *The Hosts* (2009)

35. Wood, G.: Living Dolls: A Magical History of the Quest for Mechanical Life, p. 22. Faber, London (2002)

36. Voskuhl, A.: Androids in the Enlightenment: Mechanics, Artisans, and Cultures of the Self. University of Chicago Press, Chicago (2013)

37. Miller, Z.: http://time.com/75642/obama-japan-robot-asimo/. Accessed 14 June 2015

38. Willis, M.: In direct email exchange with Marynowsky (2009)

39. Jochum, E.: IEEE International Conference on Robotics and Automation. Karlsruhe, Germany, 10 May 2013. https://docs.google.com/document/d/115qiZ6rT521bQE6IN WbHRXkeusr_9FtXrgowCle1SuM/edit. Accessed 14 June 2015

40. Turing, A.M.: Computing machinery and intelligence. Mind **59**, 433–460 (1950)

41. Weizenbaum, J.: Computer Power and Human Reason: From Judgement to Calculation. WH Freeman, San Francisco (1976)

42. MacKinlay, D.: The discreet charm of the bourgeoisie robot. RealTime (2009). http://www.realtimearts.net/studio-artist/The_Discreet_Charm_Of_The_Bourgeoisie_Robot. Accessed 14 June 2015

43. Acconci, V.: Following Piece (1969)

44. Salter, C.: Entangled: Technology and the Transformation of Performance. MIT Press, Cambridge (2010)

45. Machover, T.: (2015). http://opera.media.mit.edu/. Accessed 14 June (2015)

The Performance of Creative Machines

Petra Gemeinboeck[1(✉)] and Rob Saunders[2]

[1] Creative Robotics Lab, National Institute for Experimental Arts,
Faculty of Art and Design, University of New South Wales,
Sydney, NSW, Australia
`petra@unsw.edu.au`
[2] Design Lab, Faculty of Architecture, Design and Planning,
University of Sydney, Sydney, NSW, Australia
`rob.saunders@sydney.edu.au`

Abstract. Cybernetic and robotic agents have long played an instrumental role in the production of 'machine creativity' as a cultural discourse. This paper traces the cultural legacy of the performance of automata and discusses historical and contemporary works to explore machine creativity as a cultural, bodily practice. Creative machines are explored as performers, capable to expand the script they are given by their human creator and skillful in bidding for the audience's attention.

Keywords: Agency · Automata · Computational creativity · Performance · Robot culture

1 Introduction: Creative Machines as a Cultural Discourse

When we talk about robots, we often implicitly refer to the cultural phenomena that give form to mechanical golems, artificial pets and a cheeky, beeping can-shaped repair droid. Robots play an important role in probing, questioning and daring our relationships with machines. For Chris Csikszentmihalyi, robotics in the 21st century is "part of a dense stew of research, design, pop culture, commodity production, and fetishism" [6]. So far, according to Csikszentmihalyi, "this cultural legacy of the robot/automaton far outweighs its ostensible practical use in warfare, space exploration, or housekeeping." While labour and power (to surveil, govern, kill, etc.) are at the centre of this more or less fictional struggle, intelligence and creativity are the stimulants. Machine intelligence is not only the final frontier but has fuelled the cultural narrative of robotics long before the earliest attempts to engineer an Artificial Intelligence. Creativity as an essential ingredient of machine intelligence was listed as one of the seven grand challenges in the groundbreaking *1956 Dartmouth Proposal for Artificial Intelligence* [17].

A machine's creativity is commonly discussed in relation to anthropocentric projections of creative abilities and attributes [3,22] or based on a comparison of its outputs to human creations [12], similar to the Turing Test [20,24]. In this

© Springer International Publishing Switzerland 2016
J.T.K.V. Koh et al. (Eds.): Cultural Robotics 2015, LNAI 9549, pp. 159–172, 2016.
DOI: 10.1007/978-3-319-42945-8_13

paper we offer a different path to exploring machine creativity and its cultural potential by looking at a machine's agency through the lens of performance. Applying the concept of performance foregrounds how machines' behaviours are culturally coded and part of a network of interactions with other social agents, their immediate environment and the cultural context itself. It heightens the performance of artificial embodied agents as a bodily practice that produces cultural meanings by 'translating' software scripts "into an 'experienceable' reality" [8]. While this pertains to all robotic agents, whether deployed in industry, research or an artistic context, this paper focuses on the potential for artistic robots to extend the script given by their human creators. Capable of being sensitive to their environment and the effects they produce, their performance evolves beyond what has been set in motion by the artist.

2   The Performance of the Automaton

Since the beginning of Artificial Intelligence (AI) in the 1950s, our idea of machines has expanded from questions of instrumentality to, as Suchman argues, "include a discourse of machine as acting and interacting other" [35]. From a cultural point of view, it could be argued that machines have always been imagined and regarded as 'acting other' and, more so, not seldom have been attributed a spirit. Automated, self-moving machines have been part of human culture since ancient times. Limited to a number of pre-programmed movements, automata derive their evocative power from the skilful embodiment of their 'program'. The inanimate machine acts as if imbued with life, "[w]hat normal representative images only threaten to do, namely come alive, the automaton seems to actually realize" [13]. The automaton's deep cultural entanglement is reflected in the varying levels of "amusement, fascination, unease, and horror at the object ... in accordance to the beliefs, concerns, and needs of each period" [13]. Whether magical, eerie or exposing their machinic nature, they are considered the forerunners of today's robots.

Ancient, elegant, programmable self-propelled machine theatres have been traced back to the 1th century, with references to earlier examples from 200 B.C.E. [31]. In the 15th Century, Leonardo da Vinci realised cunningly life-like movements using irregularly shaped cams and a linking rod to push or pull the automaton's appendages. Whereas these earlier self-moving machines seemed to be driven by mysterious, magical powers, in the wake of the enlightenment, the relationship between man and machine became more complicated. Serving as "the central emblem of the entire mechanistic worldview that was dominant in the period" [13], in this era, the marvellous automaton originated from the same mindset as mechanised labour and factories that configured humans and machines to form an "organic unity" [15]. Vaucanson, the inventor of the first fully automated loom, created three famous automata: the pipe and tabor player (1737), the flute player (1738), and the digesting duck (1738). The flute player brought to life Antoine Coysevox's "Faun playing the Flute" (1709), a statue at the Jardin des Tuileries, by imbuing it with an astonishing animated anatomy.

It literally played the flute using its a mechanical lungs, tongue, lips and fingers. Exposed to view, the automaton performed its mechanical anatomy as much as its virtuous play. In *An Account of the Mechanism of an Automaton*, Vaucanson states that his "[d]esign being rather to demonstrate the Manner of the Actions, than to shew a Machine" [36].

The performance of these machines is often over-looked. Jessica Riskin talks of scientific performances and how they relied on "displaying hidden properties and principle as striking as possible" [26] to not only make them accessible but also theatrically engaging. Yet, skilfully automated scientific entertainers, such as Vaucanson's flute player, also performed the coming to life of a sculpture, the artistry of a musician, and, last but not least, a scientific model of the human body. While automata of this era are often discussed as 'simulating' life, that is, as experimental models for studying properties of natural subjects [25], we argue that they also performed the organic and its mechanisation. Straddling "the edge of life" [33] and seeking to "mechanize the passions" [29], their ability to graciously perform life, emotions and art underpins the cultural dimension and affective potential of automata[1]. The conception of creativity in the 18th century differed greatly from our contemporary understanding [11] which might explain why these automatons weren't admired for their mechanized creative acts. From a contemporary viewpoint, the breathing automated musician, pursing its mechanical lips to play the flute with the subtle nuances of a human musician [33] could easily be considered a (machinic) bodily practice that is embedded in but also produces cultural meanings.

The automaton's mechanical performance is very similar to that of many contemporary robots, whether performing their daily routine in an automated assembly line or drawing gallery visitors into their theatrical, pre-scripted performances. Most robotic artworks perform a sense of life, intelligence or other agencies uneasily attributed to the non-living through an entirely pre-programmed set of movements and behaviours. There is, however, a smaller number of works, in which the machine operates in an open loop, sensitive to its environment and other agents and capable of adapting in response. In the following, we will explore this expanded notion of machine performance, one in which creative faculty is not only persuasively mimicked [3] but materialises from the robot's ability to interact, learn and enact agency.

3   Machine Creativity

Creativity is notoriously difficult to define and the multitude of attempts shows that our understanding of creativity always is culturally situated. Its characteristics and modes of assessment have been widely discussed by researchers

---

[1] Interestingly, Vaucanson's Pipe and Tabor Player (1737), advertised in the London Magazine as "outdoing all [human] Performers on the Instrument" (see *London Magazine, or the Gentleman's Intelligencer*, vol. 13), performed early notions of superior, machinic agency, rather than human virtuosity.

in Psychology [14,27] as well as in the AI subfield of Computational Creativity [4,12,28]. The question as to whether machines can be creative is not only a complex, often contentious, philosophical issue but also has become an instrument of the scientific query into the nature of creativity (see [2]). In many ways, the very notion of a creative machine is a cultural construct[2] and whether a machine can be considered creative is much more likely a cultural judgment than a scientific finding [2]. Computational Creativity is concerned with, according to the most quoted definition, developing software "that exhibits behaviour that would be deemed creative in humans" [4], whereby, most commonly, the machine's behaviours or outputs are assessed by human experts in the respective domains [12]. Emily Howell's musical compositions, for example, have been praised to have a quality indistinguishable from human works, and yet Emily is a software-based 'composer', developed by David Cope, with the ability to learn from, and expand upon, existing works [5].

3.1   Machine Performance and Agency

If we look at creative machines as a cultural trope and practice, performance provides a useful lens to explore their nonhuman agency and its affective potential. Performance is used in a double sense here: on the one hand it reimagines the machines' acting as both a cultural and embodied practice; and on the other, it directly refers to how the machines act out their script and interact with the world. With regards to creative machines, we need to distinguish between machines that act creatively but can only follow a predetermined script and machines that can sense, learn and adapt and whose performance is open to change. There's a difference between an audience member projecting creative agency onto a robot, a robot designed to perform as if it were creative (pretending)[3], and a robot capable of extending its given script by learning to be sensitive to the effects it produces. The latter is not about assigning genuine creative capabilities to the robot, but acknowledges a robot's potential to expand the performance envelope designed by its human creator(s). An open system as such, from a performance point of view, can be looked at as capable of negotiating its environment; its performance is not entirely pre-scripted or reactive.

---

[2] In this paper, we often refer to creative machines, rather than creative robots. From a cultural viewpoint, the term 'machine' is less readily associated with humanoid forms. Creative machines, thus, open up the image of the creative robot to include more complex understandings of the machine as assemblage, always in interaction with other assemblages, including the environment, humans, the cultural context, history, etc.

[3] Cleland argues that "[f]or a robot, [successful acting] is the ability to persuasively simulate or pass as human or alive or intelligent" [3]. Following this argument, if the aim is for a robot to appear creative, its successful performance would be to persuasively simulate creative behaviour, e.g., painting robots. As we have suggested earlier, however, a pre-programmed automaton is capable of delivering such a persuasive performance; it doesn't require the advanced capabilities of a robot.

A machine's ability to extend its script is a question of agency. All machine agents are cultural actors [21], whether they are limited to operate in a closed loop or not. A machine's potential to act and affect that is defined by the audience projecting their knowledge onto the machine or the creators script, however, locates the machine's agency solely within human culture. In contrast, an open, creative machine performer materialises agency as distributed and enacted across human and nonhuman domains. One of the most fundamental differences between software-based agents and robots is that the latter are embodied; they act and share the world *with* us, in bodily ways. Without disregarding their differences, both human and nonhuman agents adapt and know because they act as part of the world [1]. We argue that, open, adaptive, embodied systems are able to take part in this negotiation, beyond their creator's intent; they perform beyond representation and actively participate in the production and distribution of cultural agency.

4   Creative Machine Performers

Andrew Pickering talks about artworks that foreground a performative rather than a representational, epistemic "aspect of being in the world" as "ontological theatre" [23]. Ontology is about, in Pickering's words, "what sorts of things there are in the world, and how they relate to one another" [23]. Following this conception, performance, particularly if beyond representation, is an ontological practice, and engaging in a creative machine's performance can be looked at as a dynamic dramaturgy of human and nonhuman agents interacting without a given script but with all the emergent possibilities this may produce. One such ontological theatre that Pickering refers to is Gordon Pask's Colloquy of Mobiles, and this pioneering work also excellently serves to materially perform the points about creative machines we have made so far.

4.1   The Colloquy of Mobiles

Gordon Pask was a major figure in British cybernetics after the 2nd World War and, perhaps lesser known, a pioneering artist and theatre designer. His work, *The Colloquy of Mobiles*, was shown at the *Cybernetic Serendipity* exhibition, curated by Jasia Reichardt at the ICA, London, in 1968. The robotic sculpture performed a dynamically evolving mating scenario between five 'mobiles'—three female robots, with soft fibreglass shapes, and two male robots, made of aluminium rectangles, Fig. 1. The work introduced machinic attributes that even today would sound advanced to museum audiences, including agency, communication, interactivity, intelligence and ability to learn. *The Colloquy of Mobiles* physically embodied Pask's cybernetic concept of the Conversation Theory [30], which he developed in tandem with his material, aesthetic experiments [9]. Yet the work is in many ways as much a humorous, social observation of humans and their nonhuman counterparts as it is a technological achievement. Pickering's [23] description of this complex work is worth quoting in full:

Fig. 1. Gordon Pask's *The Colloquy of Mobiles* (image courtesy of Amanda Heitler)

The mobiles [...] were complicated electro-mechanical robots, designated male and female, which communicated with one another via lights and sounds, and engaged in uncertain and complicated matings. The males would emit light beams, which the females would try to reflect back at them. When the reflected beam struck a particular spot on the lower parts of the males, they would be "satisfied" and go quiescentuntil their drives started to build up again. The females, too, had drives they sought to satisfy, and were adaptive in the sense that they could learn to identify individual males and remember their peculiarities.

Pickering's description emphasises the robots' performance of their drives, how they move and reconfigure themselves to fulfil them, and learn to select and adapt to particular sensations. While their behaviours evolved based on their own inner dynamics and interactions with the other robots, they were also open to outside interference. Visitors were keen to interact when they discovered that they could use mirrors and flashlights [37] to participate in this strange mating ritual (albeit only on machinic terms). *The Colloquy of Mobiles'* self-driven, dynamic performance doesn't create an artefact, but produces an endlessly emergent cycle of relations, meanings and desires; a conversation across nonhumans and humans. In Pask's own words, 'an aesthetically potent environment should [...] respond to a man, engage him in conversation and adapt its characteristics to the prevailing mode of discourse' [19].

5 From Human–Machine Performance to Machine–Machine Performance

Pask's 'conversational machines' explore the emergence of unique interaction protocols between humans and machines and between machines themselves. In the following, we look at more recent works investigating similarly emergent forms of interaction.

5.1 Performative Ecologies

Ruairi Glynn's *Performative Ecologies: Dancers* is a conversational environment, involving human and robotic agents in a dialogue using simple gestural forms [10]. In this installation, the Dancers are robots suspended in space by threads and capable of performing 'gestures' through twisting movements, Fig. 2. The fitness of their gestures is evaluated as a function of audience attention[4], independently determined by each robot through face tracking. Audience members are able to directly participate in the evolution of the machine performance by manually manipulating the robots, twisting them to record new gestures.

Fig. 2. Ruairi Glynn's *Performative Ecologies: Dancers* (image courtesy of Ruairi Glynn)

In a way, the audience is invited to physically choreograph the machines' dance in order to expand the dancers' gene pool of gestures to generate new dance sequences. The robots collaborate with each other by sharing their most

---

[4] Jon McCormacks' sonic ecosystem *Eden* uses a similar attention-based reward system to drive the musical performance of artificial agents [18].

successful moves. That is, gestures that attract the most audience attention are shared between the robots over a wireless network. Glynn's work directly links the creative act of producing new gestures with their attention-seeking performance. At the same time, the audience's attention serves to evaluate the dancers' new creations. Thus, while from the audience perspective it may appear as if these creative machine dancers perform for them, the robots in fact elicit the audience to perform with them in order to expand their dance repertoire.

5.2   The New Artist

The New Artist is an artwork with the objective to create purely robotic art— art created and performed by robots for a robotic audience [32]. The project is a collaboration of Ben Brown, Geoff Gordon, Sue Ann Hong, Marek Michalowski, Paul Scerri, Axel Straschnoy, Iheanyi Umez-Eronini, and Garth Zeglin, and is produced by Piritta Puhto. A significant part of the research project included discussions with roboticists, neurobiologists, philosophers, theatre directors, and artists to examine what 'robotic art' could be and why we would want robots to be creative performers and appreciative audiences. Some of the researchers questioned the validity of the enterprise, arguing that there is no reason for robots to make art for other robots. While others considered it to be part of a natural progression in creative development:

> "We started out with human art for humans, then we can think about machine art for humans, or human art for machines. But will we reach a point where there's machine art for machines, and humans don't even understand what they are doing or why they even like it." [34]

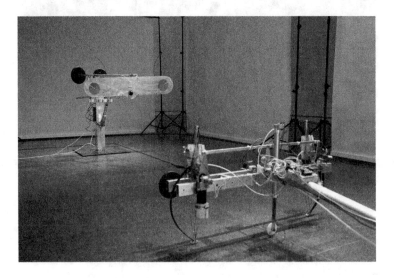

Fig. 3. *The New Artist* (image courtesy of Garth Zeglin)

The resulting work opens up a radically nonhuman view onto machine creativity by developing a performance scenario that only involves a machine artist and a machine audience member. The artist seeks to entertain the audience member by continuously evolving its performance, while the audience's role is to follow the artist's movements and express its appreciation, Fig. 3. Interestingly, this intimate scenario remains exclusive to machines and is not presented as a spectacle for humans. The installation of the work includes four opaque screens that form the performance arena but at the same time shield the robots from the prying eyes of the human audience. Thus *The New Artist* as an artwork performs the idea of a robot culture without humans—machine artists making art for machines.

5.3 Accomplice

Accomplice is a large-scale robotic installation by Petra Gemeinboeck and Rob Saunders that embeds a group of autonomous robots into the walls of a gallery. The work explores machines as self-motivated, intelligent entities and, nestled into the close contact surface of our built environment, confronts audiences with the machines' alien, social expressions. Each robotic wall-inhabitant is equipped with a punch and a camera, which they use to interact with their surrounds. They are programmed to be 'curious agents', driven to explore their world and discover things (visual patterns) they didn't expect. With a punch 'at hand', they are able to introduce changes to their world, whenever it seems already too familiar and they get 'bored'[5]. The punch enables the robots to sculpt their immediate environment by piercing wholes that eventually cause the wall to break open. Moving along the wall they share, they also use their punch to develop rhythmic knocking signals to communicate their presence to each other. As a result of this ongoing piercing and signalling activity, the walls are increasingly marked with configurations of cracks and patterns that trace the machines' appetite for change, Fig. 4.

The robots' performance is shaped by their curious disposition; the drive to seek novel 'experiences' expands their performance envelope beyond what they have been programmed to do. To these curious machines, learning and adapting are not goal driven but are based on what they discover and interpret as 'interesting' [28]. The material coupling with the wall contributes to the unpredictable evolution of the performance as the seemingly passive wall resists or accelerates the machines' eager work.

Accomplice acts out a slice of our machinic ecology—a dynamic co-mingling of processes, matter, beings and things, while foregrounding the affective potential of nonhuman, socially behaving, intelligent agents. The work is staged in

---

[5] Gordon Pask already developed an ambitious architectural machine that would reconfigure itself when it got 'bored' for the Fun Palace project, a collaboration with architect Cedric Price. Unfortunately the project was never realised. More details on the Fun Palace and Gordon Pask's pioneering concept for its dynamic architectural machine can be found in [16].

Fig. 4. Accomplice, installed at NAMOC, 2015, (image courtesy of Petra Gemeinboeck and Rob Saunders)

an unusual way in that the robots are hidden, at least at first, behind—what audiences believe to be—an existing wall. Similarly to *The New Artist*, we were not interested in creating a machine spectacle. Rather, robotics is deployed as a medium of intervention to shift the focus from representation to the machines' performative agency. While the audience plays a part in the work's wider ecology, the robots don't necessarily respond to or perform for them. This is a conception of interaction that, in Simon Penny's words, "has been expanded beyond user-machine, to larger ideas of behaviour between machines and machine systems, and between machine systems and the world" [22]. The work's affective potential is thus not in the dynamic feedback loop between the robots and the audience, but rather in their haunting physical presence and the allien-ness and unpredictability of their behaviours.

6   Discussion

In this final section, we will take a closer look at creative machine performers as cultural participants. Robots as cultural artefacts [21] are commonly considered as being constructed by a human creator, who is situated within a specific cultural context, and re-constructed by the audience, again within a given cultural context, as they project their ideas of intelligence, social agency, creativity, etc. onto the robot. Put differently, a robot is a cultural construct, with much of *what it is* and *what it can do* arising from the human cultural environment, rather than the robot itself. The argument we present in this paper, in a way, identifies a loophole for embodied, open systems—those that we consider creative machines—while acknowledging that to-date no machine exhibits creativity as it is currently recognised by humans.

As discussed earlier, open, adaptive machine systems can be considered creative machines that actively participate in an interactive conversation or an open-ended narrative, due to their capacity to adapt and learn and, thus, extend their script beyond what they have been given by their human creator. As performers, they are capable of improvising as they become increasingly sensitive to the effects they produce. Following this line of thought, they can also be considered active participants in the creation of culture, and—this is the loophole—they do so from inside their machinic context, thus extending the questions of *what a robot is* and *what a robot can do* to the nonhuman cultural domain. Creative machines, operating across human and nonhuman domains, thus provide us with a glimpse into nonhuman robot culture.

What distinguishes the creative machines discussed here from machinic agents commonly considered as being creative is that they don't produce cultural artefacts in the traditional sense. Rather, their cultural contribution is their nonhuman performance. Importantly, they don't perform for us but rather with us (or in the case of *The New Artist* neither for us nor with us). As an ontological theatre [23], the here discussed creative machine performances provide a stage for playing out different scenarios and relations across human-nonhuman cultural domains. All four works comprise more than one individual actor and present a small-scale machinic ecology with machines interacting and learning with each other.

Three of the works create machinic ecologies open to human invention. Pask's *The Colloquy of Mobiles* and Glynn's *Performative Ecologies* both invite social interactions with humans, albeit they also evolve independent of human input. It would seem that the robots in *The Colloquy of Mobiles*, playing out a machinic mating scenario, actually get on better without human interference. Yet, only people attempting to take part in trans-species mating, can experience the openness of the system without having a deeper insight into the machines' learning abilities. The Performative Ecologies, on the other hand, are actively expanding their dance gene pool by interacting with the human species. *The New Artist* and *Accomplice* both perform an ecology that is more overtly insistent on acting out 'a machine's world'. Yet, while *The New Artist* does away without humans and apparently their culture altogether, *Accomplice* situates itself into a conflict zone between the two. The dramaturgical strategy of the latter is based on an independent machine world that has nestled itself—perhaps too close—into the human world. Creative machines as such contribute to culture because they are not only staging (i.e. representing) our relationship with machines but allow us to explore them and act them out together with our nonhuman co-performers.

Finally, returning to the thought of creative machines providing us with a glimpse into nonhuman culture also opens up a view onto machine learning and its cultural implications. The creative, open systems discussed here all use machine learning in some form to allow the machines to evolve their performance based on past 'experiences' to fulfil their machine desires, learn to dance with a human, entertain another machine or to turn their world into a playground. Machine learning can be said to have already become a powerful engine driving much of our so-called human culture as it mobilises big data projects, the internet of things, the stock market, Dr. Watson, and the near future of self-driving cars. Google has long bet on machine learning to give it a competitive edge in the market, with its search algorithm allowing us to navigate the vastness of the world-wide-web and its recently rolled out Google Photos using machine learning to "make memories not manage them," in the words of director Anil Sabharwal [7]. Amongst the computer savvy, 'machine learning' has become as much a household phrase as 'artificial intelligence'.

Creative machine performers provide a different, more intimate window into machine learning and its cultural potential. They bring learning machines into our messy, embodied world and create scenarios in which we can encounter the sometimes whimsical, other times creepy, and often playful nonhuman. Here, machine learning activates a cultural, bodily practice mobilising subjective, embodied, culturally embedded experiences, rather than vast quantities of data. Perhaps their most significant cultural contribution is to open up a space for us to ask what a creative machine is. Or what it could be. Or whether we would want to live with one. The space for creative machines to exist is still an utterly human one. Just as well, given how unlikely it is that we will be able to recognise genuine nonhuman creativity.

References

1. Barad, K.: Posthumanist performativity: toward an understanding of how matter comes to matter. Signs **28**(3), 801–831 (2003)
2. Boden, M.A.: Could a robot be creative–and would we know? In: Ford, K.M., Glymour, C., Hayes, P.J. (eds.) Android Epistemology. MIT Press, Cambridge (1995)
3. Cleland, K.: Robots as social actors: audience perception of agency, emotion and intentionality in robotic performers. In: 17th International Symposium on Electronic Art (ISEA 2011). Sabanci University, Istanbul (2012). https://isea2011.sabanciuniv.edu/paper/robots-social-actors-audience-perception-agency-emotion-and-intentionality-robotic-performers
4. Colton, S., López de Mántaras, R., Stock, O.: Computational creativity: coming of age. AI Mag. **30**(3), 11–14 (2009)
5. Cope, D.: Computational creativity and music. In: Computational Creativity Research: Towards Creative Machines, pp. 309–326. Atlantis Press (2015)
6. Csikszentmihalyi, C.: Robots. In: Jones, C.A. (ed.) Sensorium: Embodied Experience, Technology, and Contemporary Art. MIT Press, Cambridge (2006)
7. Dignan, L.: Google i/o 2015: Google bets machine learning can create an edge with android, apps, cloud, 28 May 2015. http://www.zdnet.com/article/google-bets-machine-learning-can-create-an-edge-with-android-apps-cloud/
8. Dirksmeier, P., Helbrecht, I.: Time, non-representational theory and the "performative turn"–towards a new methodology in qualitative social research. Forum Qual. Soc. Res. **9**, 1–24 (2008)
9. Fernandez, M.: Gordon pask: cybernetic polymath. Leonardo **41**(2), 162–168 (2008)

10. Glynn, R.: Performative Ecologies: Dancers (2008). http://www.ruairiglynn.co.uk/portfolio/performative-ecologies/
11. Herissone, R.: Introduction. In: Herissone, R., Howard, A. (eds.) Concepts of Creativity in Seventeenth-Century England. Boydell and Brewer, Woodbridge (2013)
12. Jordanous, A.: A standardised procedure for evaluating creative systems: computational creativity evaluation based on what it is to be creative. Cogn. Comput. **4**(3), 246–279 (2012)
13. Kang, M.: Sublime Dreams of Living Machines: The Automaton in the European Imagination. Harvard University Press, Cambridge (2011)
14. Kaufman, J.C., Baer, J.: Beyond new and appropriate: who decides what is creative? Creativity Res. J. **24**(1), 83–91 (2012)
15. Kuskey, J.: The Body Machinic: Technology, Labor, and Mechanized Bodies in Victorian Culture. Ph.D. thesis, Syracuse University (2012)
16. Mathews, S.: The fun palace: cedric price's experiment in architecture and technology. Technoetic Arts J. Speculative Res. **3**(2), 73–91 (2005)
17. McCarthy, J., Minsky, M.L., Rochester, N., Shannon, C.E.: A proposal for the dartmouth summer research project on artificial intelligence. AI Magazine **27**(4), 12–14 (2006). 31 August 1955
18. McCormack, J.: On the evolution of sonic ecosystems. In: Komosinski, M., Adamatzky, A. (eds.) Artificial Life Models in Software, 2nd edn, pp. 393–414. Springer, Berlin (2009)
19. Pask, G.: A comment, a case history and a plan. In: Reichardt, J. (ed.) Cybernetics, Art, and Ideas, pp. 76–99. New York Graphics Society, Greenwich (1971)
20. Pease, A., Colton, S.: On impact and evaluation in computational creativity: a discussion of the turing test and an alternative proposal. In: Proceedings of the AISB Symposium on AI and Philosophy (2011)
21. Penny, S.: Agents as artworks: and agent design as artistic practice. In: Dautenhahn, K. (ed.) Human Cognition and Social Agent Technology. John Benjamins Publishing Company, Amsterdam (2000)
22. Penny, S.: Towards a performative aesthetics of interactivity. Fibreculture J. **132**, 72–108 (2011)
23. Pickering, A.: Ontological theatre: gordon pask, cybernetics, and the arts. Cybern. Hum. Knowing **14**(4), 43–57 (2007)
24. Riedl, M.O.: The lovelace 2.0 test of artificial creativity and intelligence. Preprint arXiv:1410.6142, arXiv (2014)
25. Riskin, J.: Eighteenth-century wetware. In: Bensaude-Vincent, B., Newman, W. (eds.) The Artificial and the Natural: An Evolving Polarity. MIT Press, Cambridge (2007)
26. Riskin, J.: Amusing physics. In: Science and Spectacle in the European Enlightenment, pp. 43–64. Ashgate, Burlington (2008)
27. Runco, M.A., Jaeger, G.J.: The standard definition of creativity. Creativity Res. J. **24**(1), 92–96 (2012)
28. Saunders, R.: Curious Design Agents and Artificial Creativity. Ph.D. thesis, The University of Sydney (2001)
29. Schaffer, S.: Enlightened automata. In: Clark, W., Golinski, J., Schaffer, S. (eds.) The Sciences in Enlightened Europe. Chicago University Press, Chicago (1999)
30. Scott, B.: Gordon pask's conversation theory: a domain independent constructivist model of human knowing. Found. Sci. (Special Issue on The Impact of Radical Constructivism on Science) **6**(4), 343–360 (2001)
31. Sharkey, N.: The programmable robot of ancient greece. New Sci. **2611**, 32–35 (2007)

32. Siemerink, E.: Axel straschnoy and his practical philosophy, 11 April 2013. http://www.argentinaindependent.com/the-arts/art/axel-straschnoy-and-his-practical-philosophy/
33. Sill, C.R.: A Survey of Androids and Audiences: 285 BCE to the Present Day. Master's thesis, Simon Fraser University (2000)
34. Straschnoy, A.: The New Artist (2008). http://www.the-new-artist.info/. Interview with Jeff Schneider, Associate Research Professor, Robotics Institute, Carnegie Mellon
35. Suchman, L.: Human/Machine Reconsidered. Cogn. Stud. **5**(1), 5–13 (1998)
36. Vaucanson, J.D.: An account of the mechanism of an automaton, or image playing on the German-flute: as it was presented in a memoire, to the gentlemen of the royal-academy of sciences at Paris. Translated by J.T. Desaguliers. T. Parker, London (1742)
37. Zeidner, J., Scholarios, D., Johnson, C.D.: Classification techniques for person-job matching: an illustration using the us army. Kybernetes **30**, 984–1005 (2001)

The Advent of Robotic Culture

Compressorhead: The Robot Band and Its Transmedia Storyworld

Alex Davies[1]([✉]) and Alexandra Crosby[2]

[1] University of New South Wales (Art & Design) (UNSWAD), Paddington, Australia
`alex.davies@unsw.edu.au`
[2] University of Technology Sydney (UTS), Sydney, Australia
`alexandra.crosby@uts.edu.au`

Abstract. Robot-human relationships are being developed and redefined due to the emerging cultural phenomenon of popular robot bands such as Compressorhead and Z-Machine. Our primary research interest in this paper is the ways in which robots relate to, interact with, and are perceived by humans - or in short, human-robot relationships. To this aim we have conducted a small-scale (multi-'species') ethnography in which we were participant observers in the ongoing production of both the 'onstage' and 'offstage' transmedia storyworld of the all-robot band, Compressorhead. We use Henry Jenkins's (2004, 2006, 2008) concept of 'transmedia storytelling' as a way of understanding how a storyworld that includes extensive human-robot interaction is simultaneously created by both humans and robots across multiple communication media platforms. In so doing, we argue that robots can indeed be seen as musicians, performers, and even celebrities, and therefore can be taken seriously as producers of culture.

Keywords: Robot music · Transmedia storytelling · Robot bands · Compressorhead

1 Introduction

On January 3rd 2013 a robot band called Compressorhead uploaded a video to YouTube: a rehearsal of their cover of the song 'Ace of Spades' (1980) by British heavy metal band Motörhead, performed in the workshop of a group of machine artists in Germany. The video went 'viral' launching a story of an all-robot band dubbed 'more metal' than any before it. At the time of writing, that video had accumulated nearly 7 million views, over 70,000 'likes', 16,000 YouTube channel subscribers, and had generated over 7,000 comments [1]. As other robot stories have done in a range of media forms over time, such as Karel Čapek's play *R.U.R.* (1920), the manga *Astro Boy* (1952) by Osamu Tezuka, the novel *The Iron Man* (1968) by Ted Hughes, or Fritz Lang's '*Maschinenmensch*' (or, "machine-human") in the film *Metropolis* (1927) to name a few, Compressorhead is challenging how humans think about machines.

This paper aims to understand the role of robots as cultural participants - and also producers - in the world of heavy metal music. Specifically, we consider Compressorhead as a cultural robotic entity, developing and participating in both material culture,

© Springer International Publishing Switzerland 2016
J.T.K.V. Koh et al. (Eds.): Cultural Robotics 2015, LNAI 9549, pp. 175–189, 2016.
DOI: 10.1007/978-3-319-42945-8_14

or the music and merchandising - and also in nonmaterial culture, or social values and norms. We use Jenkins's (2004) concept of 'transmedia storytelling' [2] as a way of understanding how a storyworld, such as one that explains human-robot interaction, is simultaneously created across multiple communication media platforms. In doing so, we argue that robots can indeed be musicians, performers, and even celebrities, and therefore need to be taken seriously as producers of culture. This mode of human-robot interaction is dispersed, heterogeneous, and is also generative in the sense that robots can also be understood to be 'evolving' cultures [3] in ongoing stories about their own futures.

Building on arguments made by a range of scholars interested in redefining culture in ways that better include non-humans (Latour 1999; Haraway 2008; Tsing 2012) [4–6], the broader implications of our position lie in the way stories about robots, told by humans - or even by robots themselves - can shift the relationships humans have to technology. By both simulating and countering quintessentially 'human' modes of performance, the robot band Compressorhead challenges two key conventions of thinking about robots that exist in public discourse: firstly the idea that robots are here to serve humans, and secondly the idea that cultural production is limited to humans. A broader, more nuanced telling of the way technological assemblages produce story-worlds in a transmedia way, as is demonstrated by Compressorhead's rise to fame, can help show that these ways of thinking belong to a human-centered past.

By exploring the storyworld of Compressorhead as it is generated across media platforms by a network of managers, venues, roadies, journalists, photographers, and groupies, as well as software, hardware, venues, and even seasons and musical fashions, we argue for an approach to cultural robotics that includes human and non-human actors in a transmedia sense. We explain how the (very human) formula (and tropes) of 'a rock band' enables a convincing image of autonomous robotic musicians, as their human developers fade (or even deliberately disappear) into the background. This explanation takes into consideration how the familiar narrative of the metal band is purposively exploited in order to suspend the disbelief of the audience and fans, until such suspension is no longer necessary, and the robots have also ostensibly become a 'band' in the human sense.

Throughout this paper we refer to Compressorhead as an 'all-robot band', as distinct from hybrid human-robot bands[1]. By this, we do not mean to imply that Compressorhead is fully autonomous; in its current iteration, the band relies heavily on its so-called 'meatbag' (i.e., human) creators. However, in its performance, and also therefore in its storytelling, it appears that the band is no more reliant on humans than is a regular human band reliant on other supportive humans to make gigs or recordings 'happen'. Compressorhead's autonomy is an important fiction, a narrative carefully and purposively created and maintained around the band. In our study of Compressorhead, we note that while the band itself is not fictional, various fictions related to the autonomy of the band-member robots are supported by a broader transmedia narrative. This narrative is generated, or is 'written', by an assemblage, composed of both humans and also non-humans.

---

[1] One example of such a human-robot (hybrid) band is: *Captured! By Robots* (1997-present). See: http://www.capturedbyrobots.com/.

Our primary research interest is focused on the ways in which robots themselves relate to, interact with, and are perceived by humans in, the world – or, human-robot relationships [7]. To this aim we have conducted a small-scale ethnography in which we were participant observers in the ongoing production of both the 'online' and 'offline' storyworld of Compressorhead. Whether this kind of ethnography is even possible is one of the questions raised by this paper, and has intersections with the emerging fields of cultural robotics, and multi-species ethnography [8]. One of the authors attended the 2013 *Big Day Out* music event in Sydney, Australia, as a backstage guest of the band Compressorhead. There, moving between the front-of-stage crowd and backstage, he observed the performance of the band and the behavior of the audience and crew. He also photographed and interviewed robot band members and their human 'minders'. Following this event, we also analyzed the official website of the band [9], their Face-Book site [10], and a range of YouTube and other online media channels, including user comments and editorial reviews of the band. In this paper we discuss in detail two published interviews with the band members, purposively selected for the way in which the journalists in each case maintained the fiction of the all-robot band.

This paper is structured in three parts. Firstly we provide a brief background to the production of popular music by robots, and an introduction to our use of the concept of transmedia storytelling. We then outline the 'origin story' of the band Compressorhead. We then move to a discussion of the 'narrative additives' we have identified which contribute to the storyworld of the 'all-robot band' fiction; these are divided into 'onstage' and 'offstage' elements. Finally we explore some of the possible future implications of Compressorhead and bands like it, for human-robot relationships.

2   Background

2.1   Context of Robot Music

Robot bands, for the purpose of this paper, can be defined as assemblages of anthropomorphic robots that perform to live audiences, and whose principal members play the instruments that typically form the foundation of contemporary bands such a guitars, bass, and drums. Since this definition is based on performance, it is important to acknowledge the rich history of robot music, of which robot bands are a subset.

Prior to the advent of recording technology such as Edison's 1877 phonograph, musical automatons were the only means to accurately and conveniently reproduce a musical performance [11]. Since these early mechanical sound machines, artists, scientists and engineers have continued development resulting in an expansive range of implementations.

Ajay Kapor defines a contemporary robotic musical instrument as "a sound-making device that automatically creates music with the use of mechanical parts, such as motors, solenoids and gears" [12]. These come in a multitude of physical manifestations ranging from abstracted machines such as Matt Heckert's 'Rotolyn' from the Mechanical Sound Orchestra (1988) [13] to Chico MacMurtrie's anthropomorphized 'Robot String Body' from *The Robotic Opera* (1992) [14]. The configuration of these musical robots ranges from individual performers such as the piano-playing WABOT-2 (1980) [15] to robot orchestras or ensembles such as *The Man and Machine Robot Orchestra* at Logos [16].

Whilst a significant number of music machines have been developed in the preceding decades, either as solo instruments or ensembles, few artists and engineers have built anthropomorphic robot bands that emulate characteristics of their human counterparts. Notable examples include the The Trons, built from discarded electronics and suburban scrap in New Zealand (2000-present) [17], the slick Japanese robot band Z-Machines (2013- present) [18], and Compressorhead (2013-Present).

2.2   A Transmedia Storyworld

It is clear that music has been an important site of experimentation for exploring whether, as (Samani *et al*, 2013) note, culture can be "not only attributed to humans, but also encompasses the cultural exchanges between robots, robots and humans, as well as other intellectual and emotional identities" [19]. As media technologies such as the internet have been broadening definitions of culture over the last two decades, the possibilities of understanding robots as operating beyond single sites of production are also becoming apparent. The concept of 'transmedia storytelling' is useful in the emerging field of cultural robotics as it provides a way of understanding how a storyworld, such as one that includes human-robot interaction, is more often than not created across multiple platforms simultaneously. According to media scholar Henry Jenkins (2008), transmedia storytelling is a process where integral elements of a fiction are dispersed systematically across multiple delivery channels for the purpose of creating a unified and coordinated entertainment experience [20]. We apply this idea to the case of Compressorhead by considering the fiction that the band is 'all robot', operating, performing and touring in a world where robots enlist the help of humans, rather than the other way around. This fiction is arguably a part of the band's appeal and popularity, and is generated and circulated through both onstage performance, and offstage mediation.

The fiction is most clear in the representation of the relationship between the robots and their human counterparts, as the band's guitar player 'Fingers' states, talking about the band's collaboration with well known musician John Wright "… We enlisted the help of John, to write new material and to facilitate our quest for rock" [21]. Additional enlisted help takes both human and non-human forms: roadies, software, technologies of distribution, designers, technicians, brands, other robots such as moving head lighting fixtures, mobile devices, and other machines.

In our review of literature relating to the transmedia framing of storyworlds we have noted that, while many of the 'stories' used as examples of transmedia depict robots - *The Matrix* cycle of movies/games/media being the most frequently cited example - the participants in the production of such storyworlds in culture are generally limited to humans. It is one goal of this essay to test this limitation. In extending transmedia storytelling to robot and human interactions, we consider Compressorhead's (human) fans, and also the many collaborators (both human and non-human) in its storyworld production.

In defining 'transmedia', Jenkins (2006) looks at the ways cultural elements create a story 'world' or 'universe' in which more varieties of narrative, and also audience interaction and engagement, are possible:

> Transmedia storytelling is the art of world making. To fully experience any fictional world, consumers must assume the role of hunters and gatherers, chasing down bits of the story across media channels, comparing notes with each other via online discussion groups, and collaborating to ensure that everyone who invests time and effort will come away with a richer entertainment experience [22].

While acknowledging that transmedia has probably always existed, Jenkins points to the way convergent digital media spaces are providing new kinds of opportunities for storyworlds. Transmedia researcher and creator Christy Dena furthers this idea in her 'transhistorical perspective' arguing that the genealogy of transmedia storytelling is complex, and should include narrative theory and multimodal discourse analysis, amongst other branches [23]. However these various elements are understood to have emerged, it is the convergent digital media spaces, in combination with live performance that are arguably at the core of Compressorhead's appeal. It is within the transmedia storyworld created by these narrative 'additives' that hidden 'clues' can be found, character histories revealed, and possible meanings and futures imagined [24]."

We do not intend to propose that transmedia is the only or definitive way to think about the multiple practices, sites of production and audiences that contribute to the Compressorhead project, or indeed any robot band for that matter. Rather our intention is to use transmedia as a lens to understand the creation of culture that crosses human-machine lines. As Dena argues "the nature and breadth of transmedia practice has been obscured because investigations have been specific to certain industries, artistic sectors and forms [25]. Robot bands have not as yet been part of those investigations, and so by including them, we hope to expand the notion of transmedia.

While there are many (and sometimes, conflicting) takes on transmedia, in this paper the focus is on the creation of storyworlds through narrative additives, in order to ask: How might transmedia explain the relationship between humans and robots in the production of an all-robot band?

3   Compressorhead

In this section we apply the idea of a transmedia storyworld to the fiction of an autonomous robot band, by outlining the ways that Compressorhead is created, across media and performance modes. We look firstly at the story of the band's origins, then at onstage performances, and lastly the band's online presence.

These sections are not intended to make a clear distinction between onstage and offstage performance by referring to the former as 'live' and the latter as not. Rather we draw attention to the ways that technological mediations work at both 'sites' (online and offline) to create the fiction of Compressorhead's autonomy from humans. The proposition that 'live' music is not clearly defined has been explored by Cultural Studies theorists elsewhere; Holt (2010) for instance points to the way that the *idea* of live music has always been a product of broad social and cultural transformation, "born in the nexus of commerce, media and entertainment" [26]. Live music, Holt argues, has only emerged as a category alongside the technologies of mass media broadcasting and recording. He points out that while 'live' came to generically refer to performance not reproduced in a studio, it also is used for technologically mediated performances: 'live show', 'live recording', 'live interaction' etc. "The live performance is associated with co-presence in the here and now, and the strict meaning involves a face-to-face relation in the same physical space [26]." He goes on to point out how distinctions in everyday language between a live recording and a studio recording or between a performance and a video have become complex as they refer to "different modes of production and perception with different economies and organisational contexts" [26]. The significance of this discussion to the notion of robot bands lies in the way that robots have been able to play 'live', as certain understandings of what the term 'playing live' means for a band have altered.

> When music exists as live music (that is to say, when it is associated with the discursive category of live music), the perspective is broadened from the music itself to questions about how, when and among whom the music is created, performed and heard in relation to practices of technological mediation. Only by examining live music in its communicative context can we understand its capacities in the production of authenticity, festivity and social presence [26].

Prior to Holt (2010), Frith (1996) argued that the value of live music is indeed about the narratives that people build around the music performance. The 'direct experiences' with music 'enable us to place ourselves in imaginative cultural narratives' [27]. In other words, concert narratives are stories about how people participate in culture, not just about music as an art form. Moving closer to the genre of music that includes the subject of this chapter, in his thesis on fan narratives within the punk music scene, Michael Janowski (2013) argues that "concert fan narratives are a vital post-contextual reproduction of music scene meaning, focused not on capital gain, but on social growth" [28]. Drawing on Frith, Janowski also notes that 'live concert is not simply a transitory experience, but also symbolizes what it means to be a music fan' [28].

3.1   Humble Beginnings

Compressorhead was purpose-built for performing 'live' robot music concerts. Formed through a creative collaboration between German, English and Australian artists from 2007 to 2012, the band consists of four anthropomorphic robot performers: 'Stickboy' on drums (assisted by a smaller robot, 'Stickboy Junior', on high-hat); 'Fingers' on guitar; and 'Bones' on bass. These machines were created by Frank Barnes (drum robots), Markus Kolb (guitar robot) and Miles van Dorssen (bass robot) [29].

The origin story of the band is mindfully told to and by fans, borrowing from classic 'rebellion' narratives of the metal and rock music genres. There are a number of attributes that correspond between the developmental trajectory of the robot band and similar (or, equivalent) human bands. For instance, one would be hard pressed to pick up a rock or metal band biography that did not commence with a story of humble 'working class' beginnings, rebelling against the establishment, defiance of authority (including breaking the law), and rehearsing in a garage. The Compressorhead origin story is similarly told by riffing on such tropes of classic rock music journalism. The robot band members paint this picture in an interview for *Gibson* magazine, the brand of guitar played by Fingers:

> *Bones*: My dad worked in a sausage factory, filling meatbags for meatbags. He hated it and it put me off slaving for humans. Now, the meatbags slave for me, tending my every need while I rock.
> *Stickboy*: It's my mum that taught me to rebel. She was a multi-armed painting robot working at the shipyard and liked to malfunction. On many occasions, she would cover the sides of ships she was meant to be painting with digital graffiti, and it left the meatbags scratching their heads and trying to re-programme her. She really influenced my attitude.
> *Fingers*: I was put to work as a speed typist at a young age and couldn't stand it. I stole a guitar from my meatbag boss's son and haven't looked back [30].

The early development of Compressorhead also features in YouTube videos. In what appear to be workshops, lounge rooms, and garages, the band performs cover songs, just as human bands often do when starting out, and then go on to write their own original material. While each band member has its own fictional backstory, Compressorhead's origin story in some ways exaggerates the early 'togetherness' of the band, as the three robots were in reality, built one at a time over a number of years [31]. Notably, in official online press materials, there is no visual evidence of their existence presented before the band's inception.

One online video dated April 2011, well before the band started performing, shows Fingers earnestly practicing AC/DC's 'TNT' alone [32]. The official narrative is that Stickboy was initially a solo performer; Fingers and Stickboy soon teamed up; then recruited bassist Bones in 2012, and thus the band was formed:

> After one year of solo shows, I realized that I was in a league of my own,' Stickboy says. 'I spent many days and nights searching the darkest, dirtiest scrap yards to find the ultimate trash metal band members to join me on this musical endeavor [33].

While it is obviously humans who would have had to put the three (or, four) players in the same room together (and, turned them on), in the entertainment media we analysed, there was very little public reference to the robots' creators. This absence of designers, writers, artists and machinists in the storyworld is an important aspect of the Compressorhead fiction. We also confirmed with Miles Van Dorssen that this is indeed deliberate, and an agreement between the band's creators [34].

Van Dorssen is a highly skilled creative engineer and artist who has had a long history developing performative machines, both individually and with collectives such as Triclops International [35]. Despite the creative expertise behind Compressorhead, the audience - and, the band's fan base - rarely if ever hear about Van Dorssen, nor the rest of the developers. The way the band is portrayed in the public sphere is both deliberately considered and carefully managed to ensure that the discourse focuses on the band members as a cultural entity unto themselves, and not merely as 'mechanical puppets' programmed by their human operators. Positioning the robot band members in this precise and consistent way establishes a coherent 'robot-led storyworld', and each official media artifact reinforces this characterization. In the words of Stickboy: "Really, though, we are all the same inside, right? Robot or human, all of us want the same things, really. Things like tune-ups, regular servicing and *Heino* instant-cake mix [21]."

3.2   Narrative Additives: Onstage

In 2013 Ken West, the director of the *Big Day Out* music festival booked Compressorhead to play on the main stage. During their first performance in Sydney, before embarking on the national festival tour, the band ascended from their 'backstage' location (using a robotic stage-riser) to perform to a capacity crowd of 57,000 people (Fig. 1).

Fig. 1. Compressorhead play the *Big Day Out* music festival, Sydney, Australia. (Image © Alex Davies 2013)

Performing at Australia's largest touring music festival, and alongside headlining acts such as The Red Hot Chili Peppers, and also The Killers, and Grinspoon contributed to the band's perceived 'authenticity' [36]. In promotional material created for the festival there was no distinction made between humans and robots; Compressorhead was billed alongside celebrities from the upper echelons of industry success rather than as a fringe robot sideshow.

During live performances, audiences applaud the band as if they are sentient humans, and as demonstrated at the 2013 *Big Day Out* Gold Coast (Australia), audiences also sing along to the songs being performed [37]. Just as some have a propensity to applaud at the end of a film in a darkened cinema, claps and whoops directed at the robot band suggest what Lombard and Ditton (1997) define as "presence as social actor within medium" [38], whereby the audience, despite being aware that the characters are not real, suppress this awareness and engage with the medium (in this case, robots) as if they were, despite it being illogical to do so. This human behavioral response to robots has also been demonstrated by Katevas et al. (2014) in the performative context of robot stand-up comedy, whereby audience members also demonstrated "laughter, applauding and enjoyment" during a humanoid robot's routine [39]. In the case of Compressorhead, fans at live performances contribute to and participate in the fiction of 'the all-robot band' by engaging with the performance in much the same way as they would for human musicians. For the spectator, the robot band members' actions conform to expectations of a human on stage; Stickboy's head logically follows the movement of his four (!)

arms as he strikes the different instruments of his drum kit, and both Fingers and Bones expressively 'rock out' with a range of bodily movements timed to the tempo or 'groove' of the current song being played. From an audience's perspective, it appears that the band is genuinely absorbed in and even enjoying the music they are performing. As research on 'a social head for a robotic musician' by Hoffman and Ju (2014) suggests, such gestures when performed by a robot band have the capacity to "communicate, engage, and offer dynamic possibilities beyond the machines' surface appearance or pragmatic motion paths" [40].

Another aspect of the onstage performance essential to the Compressorhead fiction is the appearance of error and spontaneity, which the band explains in the following way:

> [**Noisey:**] *Some might say that because you are robots, and are therefore robotic, you can't put as much passion or groove into your music as, say, Lars Ulrich does. Discuss.*
> **Bones:** You meat bags talk about passion a lot. We think passion is a polite expression for inept sloppiness, a kind of improvised error status. We sometimes error also, but it is real, and the results can be extremely expressive. Lars Ulrich can't play drums like Stickboy.
> **Stickboy:** Yes, keeping it real apparently means making lots of mistakes. We were frankly confused at first. When John asked me why I wasn't speeding up, I didn't understand. He seemed agitated and went straight to the beer fridge. Often he went straight to the beer fridge, come to think of it. We have drawn some conclusions and have adopted a well-oiled approach to our new material. Full steam ahead, as you meatbags might quip... [21]

As of this writing, Compressorhead's developers are in the process of building two additional robotic band members, a vocalist and additional guitar player [41]. Beyond the physical presence of the performers onstage, concepts also exist for additional characters that extend and reinforce the established storyworld of the band. Ideas for a robotic band manager, robotic roadies and stage-diving inflatable robots are also under discussion. Whilst not actively contributing to the musical output of the band, developments such as these would add further depth and layers to the transmedia narrative onstage, and may also provide audiences with more entry points into the fictional world. As Ian Condry (2015) notes, it is not so much the stories that are transportable across platforms, but it is the robot characters themselves that are enduring [42].

3.3 Narrative Additives: Offstage

The narrative of the all-robot metal band is told in a range of ways offstage that are arguably as important as the band's formidable onstage presence. This is indeed the case generally in the production of band culture, and Berg *et al.* (2015) have argued that it is particularly the case in heavy metal culture, where material culture (such as band shirts) play such an important role for fans in "the fusion of music, identity, and ideology" [43].

In certain highly 'participatory' channels and sites of media production, such as on YouTube, Reddit, Instagram and Twitter, it is the band's online fans that generate many narrative additives. We have analysed these sites and noticed a number of trends. Firstly, in all the comments from social media we looked at, the creators of the robots are never mentioned. In fact there is little reference to humans, or indeed, the distinction between robots and humans. This fiction, generated and maintained by fans, partly builds the Compressorhead transmedia fiction. Secondly, there is little reference to the technical aspects of the band, their machine parts, or their reliance on humans - rather, the focus of comments is typical of fans appreciating a band, and the specific music they play. Comments such as "You guys need to come to the UK please I want to meet you Stickboy" and "I want to have your babies!" are made online as if they are addressed directly to the band members [1][2]. Despite fans being aware that the band members themselves are unable to respond (and unable to procreate with humans), fans frequently use the second-person address to maintain the illusion, and participate in building the transmedia narrative. Comments that actually refer to the band members as robots are far more rare:

> "Unfortunately, like a typical drummer, he is still playing too loud for the room. The other robots kind of just accept it and try to get through the song [44]."
> "They're young. They're innocent. They're REAL Heavy Metal. And you can switch them off [44]."

While it is difficult to determine in every single case, we believe that narrative additives such as these created by fans are supported by the official Compressorhead website. The official website is the most carefully curated site of cultural production of the Compressorhead transmedia story. The site employs the traditional aesthetics of metal culture, and authenticates the storyworld as the first 'point of contact' for new fans. The site is designed as a typical band site, including candid celebrity moments, photos of individual band members during performances, and sections for 'on tour', 'bios', gig information, and press releases [9] (Fig. 2).

Fig. 2. Bones at the Hyatt NY. (© 2015 *Compressorhead* website, used with permission. Original image at: https://compressorhead.rocks/Thumbs.html, used here with permission)

---

[2] See also additional Comments at: `Compressorhead Ace of Spades', YouTube.com, 2013, https://www.youtube.com/watch?v=3RBSkq-_St8.

Perhaps as would be expected, it is beyond the official website, and in the broader internet presence of Compressorhead that we find the most powerful narrative additives, produced by human fans and even by other robots. The anonymity of some of the online content generated about the band also contributes to the idea of the 'all-robot' band, and more generally to the idea that robots could (recursively) make other robots, and contribute to (or generate) culture, including even robotic (online) fans of robot bands. A Google search (in 2015) for "compressorhead band" yields around 3,200 results, or a significant volume of online information, and discourse. How much of the online discourse is of 'human' versus 'robot' origins is also open to further investigation. Given the recent development in software-generated content in the realm of digital journalism [45], it is not a great stretch to imagine that some of this content is generated by other online robots. Software also aggregates and organizes the Google search results, for us (humans) to comprehend. As Pavlik (2013) notes: "Intelligent agents - software robots that act autonomously on behalf of another entity (typically human, but sometimes another software robot) - are also becoming increasingly common on the Web... [46]."

Compressorhead band members employ a characteristic interview style when engaging with journalists. Routinely referring to humans as 'meatbags' they consistently maintain their autonomy. The band also became further subsumed into the entertainment industry when they appeared in front of a live studio audience on the German TV program 'Die Bülent Ceylan Show', with cinematic camera conventions including crane shots and dollies enhancing the 'authenticity' of the group as rock stars, whilst the audience clapped along to AC/DC's 'TNT' played by Fingers and Stickboy [47].

From November to December 2015, the band also ran a *Kickstarter* crowdfunding campaign to fund the creation of a robot singer, and to record their debut album. Canadian songwriter and musician John Wright, of the bands NoMeansNo and The Hanson Brothers has formed a collaboration with the band to create original song material [48]. However, although the target figure of €290,000 was not reached by the funding campaign deadline, arguably the publicity around the crowdfunding campaign also extended the band's transmedia narrative, potentially increased the fan base, and increased public awareness of the band, as the campaign was also reported on various other online sites [49].

3.4 Beyond the Cover Band

Over the last decade there have been a number of advancements in the development of AI-imbued musical machines that can dynamically respond in a musical manner. For example, Festo's Sound Machines 2.0, an autonomous compositional machine comprised of five stringed instruments that respond to an input melody, and each other, to create novel musical compositions [50]. Weinburg and Driscoll (2006) describe a robotic percussionist named Haile that can "can listen to live players, analyze perceptual musical aspects in real-time, and use the product of this analysis to play along in a collaborative manner" [51], and Hoffman and Weinberg (2010) describe Shimon, a marimba playing robot with the ability to improvise in real time with a human counterpart [52]. Compressorhead also demonstrates that the sophisticated mechanics needed for robot bands to create original music content already exist. If a human musical programmer were supplanted by an AI (artificially-intelligent) counterpart, it is plausible that the next evolutionary stage of robot bands could generate truly original content by jamming with each other. Our argument here that robots can create culture will be further supported when this is the case. In the future, it now appears likely that robot bands will not only create original and valued (or, 'creative') cultural artifacts [53], but would also alter cultural perceptions of robots, due to their creative output being integrated into the transmedia fabric of the current entertainment industry.

It could also be argued that the incorporation of Compressorhead's music into remixed popular song material is more important in demonstrating that the robots are genuine cultural producers, rather than via their generation of original musical content, however that is the topic of another paper. Suffice to say that examples of remixes already exist where Compressorhead performances and recordings are treated as are any other piece of content produced by humans (or machines), losing their 'robotness' in the remix with human-produced work. One vivid example is the 2013 remix 'Hindi Rock - Punkh featuring Compressorhead (The Robot Band)' by the Indian punk band Punkh [54].

4   Conclusion

By exploring the storyworld of Compressorhead as it is generated – by both humans and robots - across media platforms and channels, we have argued for a transmedia approach to examining cultural robotics. While there is currently (and may always be) a focus on the question of whether robots can generate 'original' creative cultural artifacts, we hope that our essay has raised other and deeper questions. As robot bands continue to collaborate with humans in future, will humans continue to expand their willingness to collaborate in new ways? One way is through direct creative collaboration, as between John Wright and Compressorhead, and Squarepusher (aka Tom Jenkinson) with Z-Machine, while other modes of collaboration are more subtle. We have pointed to some of these by thinking about the ways transmedia 'narrative additives' contribute to the storyworld of the 'all-robot' band, including journalists interviewing robot band members; editors giving them media space; promoters billing the band alongside human headliners; fans rocking out to their 'live' sets; and of course, academics giving them attention as serious cultural producers. If humans are viewed as both the 'tool-making animal' and also, as 'the storytelling animal' [55], then, as Compressorhead's bass player Bones endearingly points out, robot-human relationships may become ever-increasingly fundamental, as our mutual culture evolves:

[**Leonard:**] *Which other touring bands do you like hanging out with?*
Bones: They're all meatbags and don't understand our humor. We do appreciate their abilities to rock and this is our common language, I guess [30].

References

1. Compressorhead Ace of Spades. YouTube.com (2013). https://www.youtube.com/watch?v=3RBSkq-_St8. Accessed 7 Dec 2015
2. Jenkins, H.: The cultural logic of media convergence. Int. J. Cult. Stud. **7**(1), 33–43 (2004)

3. Saadatian, E., Samani, H., Fernando, N., Polydorou, D., Pang, N., Nakatsu, R.: Towards the definition of cultural robotics. In: 2013 International Conference on Culture and Computing (Culture Computing), pp. 167–168. IEEE (2013)
4. Latour, B.: A collective of humans and nonhumans: following Daedalus's Labirynth. In: Latour's, Pandora's Hope. Essays on the Reality of Science Studies, pp. 174–215. Harvard University Press, London (1999)
5. Haraway, D.J.: When Species Meet. University of Minnesota Press, Minneapolis (2008)
6. Tsing, A.: Unruly edges: mushrooms as companion species - for donna haraway. Environ. Humanit. **1**, 141–154 (2012)
7. Geiger, R.S.: Are computers merely supporting cooperative work: towards an ethnography of bot development. In: Proceedings of the 2013 Conference on Computer Supported Cooperative Work Companion, pp. 51–56. ACM (2013)
8. Cultural Anthropology journal, Special Issue: 'Multispecies Ethnography 25(4), 545–687 (2010)
9. Compressorhead, official band website: https://compressorhead.rocks/
10. Compressorhead, official band FaceBook page: https://www.facebook.com/compressorhead/
11. Murphy, J.: Expressive Musical Robots: Building, Evaluating, and Interfacing with an Ensemble of Mechatronic Instruments, doctoral thesis, Victoria University of Wellington, New Zealand (2014)
12. Kapur, A.: A history of robotic musical instruments. In: Proceedings of the International Computer Music Conference, pp. 21–28 (2005)
13. Mechanical Sound Orchestra (1988-) by Matt Heckert. http://v2.nl/archive/works/mechanical-sound-orchestra
14. Robotic Opera (MacMurtrie 1992). http://www.amorphicrobotworks.org/works/early/
15. Roads, C.: The Tsukuba musical robot. Comput. Music J. 39–43 (1986). http://www.roboticstoday.com/robots/wabot-ii
16. Maes, L., Raes, G.W., Rogers, T.: The man and machine robot orchestra at logos. Comput. Music J. **35**(4), 28–48 (2011)
17. Steve: [user]. The Trons – Robot Band from New Zealand. Robots.Net (2010). http://robots.net/article/3079.html
18. Watercutter, A.: A Virtuoso Robot Band Whose Guitarist Has 78 Fingers. Wired.com (2014). http://www.wired.com/2014/04/squarepusher-robot-music/
19. Samani, H., Saadatian, E., Pang, N., Polydorou, D., Newton Fernando, O.N., Nakatsu, R., Valino Koh, J.T.K.: Cultural robotics: the culture of robotics and robotics in culture. Int. J. Adv. Robot. Sys. **10**(400), 2 (2013). doi:10.5772/57260
20. Jenkins, H.: Convergence Culture: Where Old and New Media Collide, pp. 97–98. New York University Press, New York (2008). (Revised Edn)
21. Shreurs, J.: Robot Band Recruits Human Member to Write Songs, Hilarity Ensues. Noisey (2015). http://noisey.vice.com/en_ca/blog/compressorhead-nomeansno-johnwright-interview
22. Jenkins, H.: Convergence Culture: Where old and new media collide, p. 21. NYU Press, New York (2006)
23. Dena, C.: Transmedia as "Unmixed Media" Aesthetics. In: Polson, D., Cooke, A.M., Velikovsky, J.T., Brackin, A. (eds) Transmedia Practice: A Collective Approach, pp. 3–16 (2014)
24. Jenkins, p. 114 (2008)
25. Dena, C. (2014)
26. Holt, F.: The economy of live music in the digital age. Eur. J. Cult. Stud. **13**, 243–261 (2010). doi:10.1177/1367549409352277. p. 245

27. Frith, S.: Music and identity. In: Hall, S., Du Gay, P. (eds.) Questions of Cultural Identity, p. 124. SAGE, London (1996)

28. Janowski, M.J.: Those who slay together, stay together: a thematic analysis of concert fan narratives and the I-57 youth punk music scene. Masters Thesis, Eastern Illinois University, Illinois (2013). http://thekeep.eiu.edu/theses/1157/

29. Compressorhead Press Kit (2015). https://compressorhead.rocks/press-kit/Compressorhead-Presskit-en.pdf

30. Leonard, M.: Meet Compressorhead - The World's Most Metal Band, Gibson.com (2013). http://www.gibson.com/news-lifestyle/features/en-us/meet-compressorhead-the-worlds-most-metal-band.aspx

31. Khanna, V.: Ep #227: John Wright & Frank Barnes of Compressorhead. Kreative Kontrol (2015). http://vishkhanna.com/2015/11/26/ep-227-john-wright-frank-barnes-of-compressorhead/

32. Fingers practicing AC/DC's "TNT": YouTube.com (2011). https://www.youtube.com/watch?v=KoZfesn6soI

33. Vega, C.: Featured Artist: Compressorhead (2015). Cerwin-Vega Professional Audio website, http://www.cerwinvega.com/pro-audio/artists-venues/cerwin-vega-featured-artist-compressorhead.html

34. Personal communications with Miles Van Dorssen (2015)

35. 'Acoustic Emitter #1' project involving Van Dorssen and Triclops International (2000). http://schizophonia.com/portfolio/acousticemitter/

36. http://www.abc.net.au/triplej/events/bigdayout/13/#item=compressorhead

37. Compressorhead: Joan Jett & the Blackhearts "I Love Rock'n Roll" (*Big Day Out*, Gold Coast), YouTibe.com (2013). https://www.youtube.com/watch?v=rFfiULkjjI4

38. Lombard, M., Ditton, T.: At the heart of it all: the concept of presence. J. Comput.-Mediated Commun. **3**(2) (1997)

39. Katevas, K., Healey, P.G.T., Harris, M.T.: Robot stand-up: engineering a comic performance. In: Proceedings of the 2014 Workshop on Humanoid Robots and Creativity, Madrid, Spain (2014). doi:10.13140/2.1.2329.6807

40. Hoffman, G., Ju, W.: Designing robots with movement in mind. J. Hum.-Robot Interact. **3**(1), 89–122 (2008)

41. Crowe, S.: Compressorhead Robot Band Launches Kickstarter for Album, Vocalist. Robotics Trends (2015). http://www.roboticstrends.com/article/compressorhead_robot_band_launches_kickstarter_for_album_vocalist. The band's plans for a robot vocalist and additional guitarist were also confirmed via personal communication

42. Condry, I.: In: Whiteacre, A. (ed.) Podcast: "Robots and Media: Science Fiction, Anime, Transmedia, and Technology". MIT, Mass (2015). http://cmsw.mit.edu/podcast-robots-and-media-scien/

43. Berg, A., Gulden, T., Hiort af Ornäs, V., Pavel, N. Sjøvoll, V.: The Metal T-Shirt: Transmedia Storytelling In Products. Paper presented at Modern Heavy Metal Conference 2015, Helsinki (2015). http://www.modernheavymetal.net/p/important-dates.html

44. User 'stay_fr0sty'. Literally heavy metal, Reddit (2013). https://www.reddit.com/r/videos/comments/15y6f7/literally_heavy_metal/?limit=500

45. Clerwall, C.: Enter the robot journalist: users' perceptions of automated content. Journalism Pract. **8**(5), 519–531 (2014)

46. Pavlik, J.V.: Journalism and New Media. Columbia University Press, New York (2013)

47. Compressorhead - Fingers and Stickboy – TNT. YouTube.com (2013). https://www.youtube.com/watch?v=qKAWBNl4MeA

48. Barnes, F., Kolb, M., Plum, S.: Building a Robot Lead Vocalist and producing first Album. Kickstarter.com (2015). https://www.kickstarter.com/projects/1335681368/building-robot-lead-vocalist-and-producing-first-a

49. Cook, J.: A German rock band made of robots is raising money on Kickstarter to build themselves a singer. Business Insider UK (2015). http://www.businessinsider.com.au/german-robot-rock-band-compressorhead-is-raising-money-on-kickstarter-to-build-a-singer-2015-11 and see: Staff Writer, 'All-Robot Band Compressorhead Launch Kickstarter To Build Vocalist For Debut LP', TheMusic.com.au, http://themusic.com.au/news/all/2015/11/03/all-robot-band-compressorhead-launch-kickstarter-to-build-vocalist-for-debut-lp/

50. Festo's Sound Machines 2.0: http://www.festo.com/net/SupportPortal/Files/156744/Brosch_FC_Soundmachines_EN_lo_L.pdf

51. Weinberg, G., Driscoll, S.: Robot-human interaction with an anthropomorphic percussionist. In: Proceedings of the SIGCHI Conference on Human Factors in Computing Systems, pp. 1229–1232. ACM (2006)

52. Hoffman, G., Weinberg, G.: Shimon: an interactive improvisational robotic marimba player. In: CHI'10 Extended Abstracts on Human Factors in Computing Systems, pp. 3097–3102. ACM (2010)

53. Runco, M.A., Jaeger, G.J.: The standard definition of creativity. Creativity Res. J. **24**(1), 92–96 (2012). http://creativity.netslova.ru/Definitions_of_creativity.html (2007)

54. Hindi Rock - Punkh featuring Compressorhead (The Robot Band). YouTube.com (2013). https://youtu.be/b5gFSkDhsKg?list=PLQ_3yhvsN8UMoIuJ-utdplvQs064EtJRv

55. Gottschall, J.: The Storytelling Animal: How Stories Make Us Human. Houghton Mifflin Harcourt, Boston (2012)

Author Index

Printed in the United States
By Bookmasters